The Lives
of a Cell

The Lives of a Cell

Notes of a
Biology Watcher

Lewis Thomas

The Viking Press
New York

Contents

The Lives of a Cell

The Lives of a Cell

W e are told that the trouble with Modern Man is that he has been trying to detach himself from nature. He sits in the topmost tiers of polymer, glass, and steel, dangling his pulsing legs, surveying at a distance the writhing life of the planet. In this scenario, Man comes on as a stupendous lethal force, and the earth is pictured as something delicate, like rising bubbles at the surface of a country pond, or flights of fragile birds.

But it is illusion to think that there is anything fragile about the life of the earth; surely this is the toughest membrane imaginable in the universe, opaque to probability, impermeable to death. We are the delicate part, transient and vulnerable as cilia. Nor is it a new thing for man to invent an existence that he imagines to be above the rest of life; this has been his most consistent intellectual exertion down the millennia. As illusion, it has never worked out to his satisfaction in the past, any more than it does today. Man is embedded in nature.

The biologic science of recent years has been making this a more urgent fact of life. The new, hard problem will be to cope with the dawning, intensifying realization of just how interlocked we are. The old, clung-to notions most of us have held about our special lordship are being deeply undermined.

Item. A good case can be made for our nonexistence as entities. We are not made up, as we had always supposed, of successively enriched packets of our own parts. We are shared, rented, occupied. At the interior of our cells, driving them, providing the oxidative energy that sends us out for the improvement of each shining day, are the mitochondria, and in a strict sense they are not ours. They turn out to be little separate creatures, the colonial posterity of migrant prokaryocytes, probably primitive bacteria that swam into ancestral precursors of our eukaryotic cells and stayed there. Ever since, they have maintained themselves and their ways, replicating in their own fashion, privately, with their own DNA and RNA quite different from ours. They are as much symbionts as the rhizobial bacteria in the roots of beans. Without them, we would not move a muscle, drum a finger, think a thought.

Mitochondria are stable and responsible lodgers, and I choose to trust them. But what of the other little animals, similarly established in my cells, sorting and balancing me, clustering me together? My centrioles, basal bodies, and probably a good many other more obscure tiny beings at work inside my cells, each with its own special genome, are as foreign, and as essential, as aphids in anthills. My cells are no longer the pure line entities I was raised with; they are ecosystems more complex than Jamaica Bay.

I like to think that they work in my interest, that each breath they draw for me, but perhaps it is they who walk through the local park in the early morning, sensing my senses, listening to my music, thinking my thoughts.

I am consoled, somewhat, by the thought that the green plants are in the same fix. They could not be plants, or green, without their chloroplasts, which run the photosynthetic enterprise and generate oxygen for the rest of us. As it turns out, chloroplasts are also separate creatures with their own genomes, speaking their own language.

We carry stores of DNA in our nuclei that may have come in, at one time or another, from the fusion of ancestral cells and the linking of ancestral organisms in symbiosis.

Our genomes are catalogues of instructions from all kinds of sources in nature, filed for all kinds of contingencies. As for me, I am grateful for differentiation and speciation, but I cannot feel as separate an entity as I did a few years ago, before I was told these things, nor, I should think, can anyone else.

Item. The uniformity of the earth's life, more astonishing than its diversity, is accountable by the high probability that we derived, originally, from some single cell, fertilized in a bolt of lightning as the earth cooled. It is from the progeny of this parent cell that we take our looks; we still share genes around, and the resemblance of the enzymes of grasses to those of whales is a family resemblance.

The viruses, instead of being single-minded agents of disease and death, now begin to look more like mobile genes. Evolution is still an infinitely long and tedious biologic game, with only the winners staying at the table, but the rules are beginning to look more flexible. We live in a dancing matrix of viruses; they dart, rather like bees, from organism to organism, from plant to insect to mammal to me and back again, and into the sea, tugging along pieces of this genome, strings of genes from that, transplanting grafts of DNA, passing around heredity as though at a great party. They may be a mechanism for keeping new, mutant kinds of DNA in the widest circulation among us. If this is true, the odd virus disease, on which we must focus so much of our attention in medicine, may be looked on as an accident, something dropped.

Item. I have been trying to think of the earth as a kind of organism, but it is no go. I cannot think of it this way. It is too big, too complex, with too many working parts lacking visible connections. The other night, driving through a hilly, wooded part of southern New England, I wondered about this. If not like an organism, what is it like, what is it *most* like? Then, satisfactorily for that moment, it came to me: it is *most* like a single cell.

Thoughts for a Countdown

There is ambiguity, and some symbolism, in the elaborate ritual observed by each returning expedition of astronauts from the moon. They celebrate first of all the inviolability of the earth, and they re-enact, each time, in stereotyped choreography, our long anxiety about the nature of life. They do not, as one might expect, fall to their knees and kiss the carrier deck; this would violate, intrude upon, contaminate the deck, the vessel, the sea around, the whole earth. Instead, they wear surgical masks. They walk briskly, arms up, untouching, into a sterile box. They wave enigmatically, gnotobiotically, to the President from behind glass panes, so as not to breathe moondust on him. They are levitated to another sealed box in Houston, to wait out their days in quarantine, while inoculated animals and tissues cultures are squinted at for omens.

It is only after the long antiseptic ceremony has been completed that they are allowed out into the sun, for the ride up Broadway.

A visitor from another planet, or another century, would view the exercise as precisely lunatic behavior, but no one from outside would understand it. We must do things this way, these days. If there should be life on the moon, we must begin by fearing it. We must guard against it, lest we catch something.

It might be a microbe, a strand of lost nucleic acid, a molecule of enzyme, or a nameless hairless little being with sharp gray eyes. Whatever, once we have imagined it, foreign and therefore hostile, it is not to be petted. It must be locked up. I imagine the debate would turn on how best to kill it.

It is remarkable that we have all accepted this, without hooting, as though it simply conformed to a law of nature. It says something about our century, our attitude toward life, our obsession with disease and death, our human chauvinism.

There are pieces of evidence that we have had it the wrong way round. Most of the associations between the living things we know about are essentially cooperative ones, symbiotic in one degree or another; when they have the look of adversaries, it is usually a standoff relation, with one party issuing signals, warnings, flagging the other off. It takes long intimacy, long and familiar interliving, before one kind of creature can cause illness in another. If there were to be life on the moon, it would have a lonely time waiting for acceptance to membership here. We do not have solitary beings. Every creature is, in some sense, connected to and dependent on the rest.

It has been estimated that we probably have real knowledge of only a small proportion of the microbes of the earth, because most of them cannot be cultivated alone. They live together in dense, interdependent communities, feeding and supporting the environment for each other, regulating the balance of populations between different species by a complex system of chemical signals. With our present technology, we can no more isolate one from the rest, and rear it alone, than we can keep a single bee from drying up like a desquamated cell when removed from his hive.

The bacteria are beginning to have the aspect of social animals; they should provide nice models for the study of interactions between forms of life at all levels. They live by collaboration, accommodation, exchange, and barter. They, and the fungi, probably with help from a communication system laid on by the viruses, comprise the parenchyma of the soil (someone has suggested that humic acid, to which the microbes contribute, is a sort of counterpart for the ground substance of our own connective tissue). They live on each other. Sometimes they live inside each other; the *Bdellovibrio* penetrate the walls of other bacteria, tuck themselves up inside, replicate, and burst out again as though they thought themselves phages. Some microbial communities extend so deeply into the affairs of higher forms of life as to seem like new kinds of tissue in plants and animals. The rhizobial bacteria that swarm over the root hairs of leguminous plants have the look of voracious, invasive pathogens, but the root nodules that they then construct, in collaboration with the plant cells, become the earth's chief organ for nitrogen fixation. The production of leghemoglobin in the membrane-lined space between plant and bacterial cells is an example of the high technology of symbiosis; the protein is synthesized by the plant, but only on instructions from the bacteria, and it is possible that the plant DNA for coding this substance came originally from the microbe, early in the evolution of the arrangement.

The bacteria that live in the tissues of insects, like those incorporated into the mycetocytes of cockroaches and termites, have the appearance of specialized organs in their hosts. It is not yet clear what they accomplish for the insect, but it is known that the species cannot survive long without them. They are transmitted, like mitochondria, from generation to generation of eggs.

It has been proposed that symbiotic linkages between prokaryotic cells were the origin of eukaryotes, and that fusion between different sorts of eukaryotes (e.g., motile, ciliated cells joined to phagocytic ones) led to the construction of the communities that eventually turned out to be metazoan creatures. If this is true, the marks of identity, distinguishing self from non-self, have long since been blurred. Today, in the symbiotic associations that dominate so much of the life of the sea, there is rarely a question of who is who, even when the combination functions like a single animal. The anemones that fasten themselves to the shells, even to the claws, of certain crabs are capable of recognizing precisely the molecular configurations that identify those surfaces: the crab, for his part, can recognize his own anemone, and will sometimes seek him out and attach him to the shell like an ornament. The damsel fish that have become, from their point of view, functioning parts of certain species of anemones adapt themselves when very young to life among the lethal tentacles of their host; they cannot just swim in forthwith—they must dart around the edges until labeled at their surfaces by markers acceptable to the anemone.

Sometimes, in the course of the modulation of relations between animals, there are inventions that seem to have been thought up on the spur of the moment, like propositions to be submitted for possible evolution. Some are good-humored, even witty. Certain Australian surf bathers, several years ago, were stung by tiny creatures that turned out to be nudibranchs armed with the stingers of Portuguese men-of-war. Having fed on jellyfish, the Glaucus community had edited their meal and allowed the stinging cells to make their way to the surface of their new host, thus creating, for the time, a sort of instant hybrid with, allowing for some asymmetry, the essential attributes of each partner.

Even when circumstances require that there be winners and losers, the transaction is not necessarily a combat. The aloofness displayed for each other by members of the marine coelenterate species of Gorgonaceae suggests that mechanisms for preserving individuality must have existed long before the evolution of immunity. The gorgonians tend to grow in closely packed, branching masses, but they do not fuse to each other; if they did, their morphogenesis would doubtless become a shambles. Theodor, in a series of elegant experiments, has shown that when two individuals of the same species are placed in close contact, the smaller of the two will always begin to disintegrate. It is autodestruction due to lytic mechanisms entirely under the governance of the smaller partner. He is not thrown out, not outgamed, not outgunned; he simply chooses to bow out. It is not necessarily a comfort to know that such things go on in biology, but it is at least an agreeable surprise.

The oxygen in the atmosphere is the exhalation of the chloroplasts living in plants (also, for our amazement, in the siphons of giant clams and lesser marine animals). It is a natural tendency for genetically unrelated cells in tissue culture to come together, ignoring species differences, and fuse to form hybrid cells. Inflammation and immunology must indeed be powerfully designed to keep us apart; without such mechanisms, involving considerable effort, we might have developed as a kind of flowing syncytium over the earth, without the morphogenesis of even a flower.

Perhaps we will find it possible to accommodate other forms of life, from other planets, out of sheer good nature. We are, after all, a planet where the rain contains vitamin B_{12}! There is enough of it, by Parker's calculation, when convective windstorms occur at the time of farmland cultivation and swirl it from the soil into the upper atmosphere, to produce a visible bloom of Euglena in a fair-sized pond.

On Societies
as Organisms

Viewed from a suitable height, the aggregating clusters of medical scientists in the bright sunlight of the boardwalk at Atlantic City, swarmed there from everywhere for the annual meetings, have the look of assemblages of social insects. There is the same vibrating, ionic movement, interrupted by the darting back and forth of jerky individuals to touch antennae and exchange small bits of information; periodically, the mass casts out, like a trout-line, a long single file unerringly toward Childs's. If the boards were not fastened down, it would not be a surprise to see them put together a nest of sorts.

It is permissible to say this sort of thing about humans. They do resemble, in their most compulsively social behavior, ants at a distance. It is, however, quite bad form in biological circles to put it the other way round, to imply that the operation of insect societies has any relation at all to human affairs. The writers of books on insect behavior generally take pains, in their prefaces, to caution that insects are like creatures from another planet, that their behavior is absolutely foreign, totally unhuman, unearthly, almost unbiological. They are more like perfectly tooled but crazy little machines, and we violate science when we try to read human meanings in their arrangements.

It is hard for a bystander not to do so. Ants are so much like human beings as to be an embarrassment. They farm fungi, raise aphids as livestock, launch armies into wars, use

11

chemical sprays to alarm and confuse enemies, capture slaves. The families of weaver ants engage in child labor, holding their larvae like shuttles to spin out the thread that sews the leaves together for their fungus gardens. They exchange information ceaselessly. They do everything but watch television.

What makes us most uncomfortable is that they, and the bees and termites and social wasps, seem to live two kinds of lives: they are individuals, going about the day's business without much evidence of thought for tomorrow, and they are at the same time component parts, cellular elements, in the huge, writhing, ruminating organism of the Hill, the nest, the hive. It is because of this aspect, I think, that we most wish for them to be something foreign. We do not like the notion that there can be collective societies with the capacity to behave like organisms. If such things exist, they can have nothing to do with us.

Still, there it is. A solitary ant, afield, cannot be considered to have much of anything on his mind; indeed, with only a few neurons strung together by fibers, he can't be imagined to have a mind at all, much less a thought. He is more like a ganglion on legs. Four ants together, or ten, encircling a dead moth on a path, begin to look more like an idea. They fumble and shove, gradually moving the food toward the Hill, but as though by blind chance. It is only when you watch the dense mass of thousands of ants, crowded together around the Hill, blackening the ground, that you begin to see the whole beast, and now you observe it thinking, planning, calculating. It is an intelligence, a kind of live computer, with crawling bits for its wits.

At a stage in the construction, twigs of a certain size are needed, and all the members forage obsessively for twigs of just this size. Later, when outer walls are to be finished, thatched, the size must change, and as though given new orders by telephone, all the workers shift the search to the new twigs. If you disturb the arrangement of a part of the Hill, hundreds of ants will set it vibrating, shifting, until it is put right again. Distant sources of food are somehow sensed, and long lines, like tentacles, reach out over the ground, up over walls, behind boulders, to fetch it in.

Termites are even more extraordinary in the way they seem to accumulate intelligence as they gather together. Two or three termites in a chamber will begin to pick up pellets and move them from place to place, but nothing comes of it; nothing is built. As more join in, they seem to reach a critical mass, a quorum, and the thinking begins. They place pellets atop pellets, then throw up columns and beautiful, curving, symmetrical arches, and the crystalline architecture of vaulted chambers is created. It is not known how they communicate with each other, how the chains of termites building one column know when to turn toward the crew on the adjacent column, or how, when the time comes, they manage the flawless joining of the arches. The stimuli that set them off at the outset, building collectively instead of shifting things about, may be pheromones released when they reach committee size. They react as if alarmed. They become agitated, excited, and then they begin working, like artists.

Bees live lives of organisms, tissues, cells, organelles, all at the same time. The single bee, out of the hive retrieving sugar (instructed by the dancer: "south-southeast for seven hundred meters, clover—mind you make corrections for the sundrift") is still as much a part of the hive as if attached by a filament. Building the hive, the workers have the look of embryonic cells organizing a developing tissue; from a distance they are like the viruses inside a cell, running off row after row of symmetrical polygons as though laying down crystals. When the time for swarming comes, and the old queen prepares to leave with her part of the population, it is as though the hive were involved in mitosis. There is an agitated moving of bees back and forth, like granules in cell sap. They distribute themselves in almost precisely equal parts, half to the departing queen, half to the new one. Thus, like an egg, the great, hairy, black and golden creature splits in two, each with an equal share of the family genome.

The phenomenon of separate animals joining up to form an organism is not unique in insects. Slime-mold cells do it all the time, of course, in each life cycle. At first they are single amebocytes swimming around, eating bacteria, aloof from each other, untouching, voting straight Republican. Then, a bell sounds, and acrasin is released by special cells toward which the others converge in stellate ranks, touch, fuse together, and construct the slug, solid as a trout. A splendid stalk is raised, with a fruiting body on top, and out of this comes the next generation of amebocytes, ready to swim across the same moist ground, solitary and ambitious.

Herring and other fish in schools are at times so closely integrated, their actions so coordinated, that they seem to be functionally a great multi-fish organism. Flocking birds, especially the seabirds nesting on the slopes of offshore islands in Newfoundland, are similarly attached, connected, synchronized.

Although we are by all odds the most social of all social animals—more interdependent, more attached to each other, more inseparable in our behavior than bees—we do not often feel our conjoined intelligence. Perhaps, however, we are linked in circuits for the storage, processing, and retrieval of information, since this appears to be the most basic and universal of all human enterprises. It may be our biological function to build a certain kind of Hill. We have access to all the information of the biosphere, arriving as elementary units in the stream of solar photons. When we have learned how these are rearranged against randomness, to make, say, springtails, quantum mechanics, and the late quartets, we may have a clearer notion how to proceed. The circuitry seems to be there, even if the current is not always on.

The system of communications used in science should provide a neat, workable model for studying mechanisms of information-building in human society. Ziman, in a recent *Nature* essay, points out, "the invention of a mechanism for the systematic publication of *fragments* of scientific work may well have been the key event in the history of modern science.". He continues:

A regular journal carries from one research worker to another the various . . . observations which are of common interest. . . . A typical scientific paper has never pretended to be more than another little piece in a larger jigsaw—not significant in itself but as an element in a grander scheme. *This technique, of soliciting many modest contributions to the store of human knowledge, has been the secret of Western science since the seventeenth century, for it achieves a corporate, collective power that is far greater than any one individual can exert* [italics mine].

With some alternation of terms, some toning down, the passage could describe the building of a termite nest.

It is fascinating that the word "explore" does not apply to the searching aspect of the activity, but has its origins in the sounds we make while engaged in it. We like to think of exploring in science as a lonely, meditative business, and so it is in the first stages, but always, sooner or later, before the enterprise reaches completion, as we explore, we call to each other, communicate, publish, send letters to the editor, present papers, cry out on finding.

A Fear of
Pheromones

What are we going to do if it turns out that we have pheromones? What on earth would we be doing with such things? With the richness of speech, and all our new devices for communication, why would we want to release odors into the air to convey information about anything? We can send notes, telephone, whisper cryptic invitations, announce the giving of parties, even bounce words off the moon and make them carom around the planets. Why a gas, or droplets of moisture made to be deposited on fence posts?

Comfort has recently reviewed the reasons for believing that we are, in fact, in possession of anatomic structures for which there is no rational explanation except as sources of pheromones—tufts of hair, strategically located apocrine glands, unaccountable areas of moisture. We even have folds of skin here and there designed for the controlled nurture of bacteria, and it is known that certain microbes eke out a living, like eighteenth-century musicians, producing chemical signals by ornamenting the products of their hosts.

Most of the known pheromones are small, simple molecules, active in extremely small concentrations. Eight or ten carbon atoms in a chain are all that are needed to generate precise, unequivocal directions about all kinds of matters—when and where to cluster in crowds, when to disperse, how to behave to the opposite sex, how to ascertain what *is* the

16

opposite sex, how to organize members of a society in the proper ranking orders of dominance, how to mark out exact boundaries of real estate, and how to establish that one is, beyond argument, one's self. Trails can be laid and followed, antagonists frightened and confused, friends attracted and enchanted.

The messages are urgent, but they may arrive, for all we know, in a fragrance of ambiguity. "At home, 4 p.m. today," says the female moth, and releases a brief explosion of bombykol, a single molecule of which will tremble the hairs of any male within miles and send him driving upwind in a confusion of ardor. But it is doubtful if he has an awareness of being caught in an aerosol of chemical attractant. On the contrary, he probably finds suddenly that it has become an excellent day, the weather remarkably bracing, the time appropriate for a bit of exercise of the old wings, a brisk turn upwind. En route, traveling the gradient of bombykol, he notes the presence of other males, heading in the same direction, all in a good mood, inclined to race for the sheer sport of it. Then, when he reaches his destination, it may seem to him the most extraordinary of coincidences, the greatest piece of luck: "Bless my soul, what have we here!"

It has been soberly calculated that if a single female moth were to release all the bombykol in her sac in a single spray, all at once, she could theoretically attract a trillion males in the instant. This is, of course, not done.

Fish make use of chemical signals for the identification of individual members of a species, and also for the announcement of changes in the status of certain individuals. A catfish that has had a career as a local leader smells one way, but as soon as he is displaced in an administrative reorganization, he smells differently, and everyone recognizes the loss of standing. A bullhead can immediately identify the water in which a recent adversary has been swimming, and he can distinguish between this fish and all others in the school.

There is some preliminary, still fragmentary evidence for important pheromones in primates. Short-chain aliphatic compounds are elaborated by female monkeys in response to estradiol, and these are of consuming interest to the males. Whether there are other sorts of social communication by pheromones among primates is not known.

The possibility that human beings are involved in this sort of thing has not attracted much attention until recently. It is still too early to say how it will come out. Perhaps we have inherited only vestiges of the organs needed, only antique and archaic traces of the fragrance, and the memory may be forever gone. We may remain safe from this new challenge to our technology, and, while the twentieth century continues to run out in concentric circles down the drain, we may be able to keep our attention concentrated on how to get energy straight from the sun.

But there are just the slightest suggestions, hints of what may be ahead. Last year it was observed that young women living at close quarters in dormitories tended to undergo spontaneous synchronization of their menstrual cycles. A paper in *Nature* reported the personal experience of an anonymous, quantitatively minded British scientist who lived for long stretches in isolation on an offshore island, and discovered, by taking the dry weight of the hairs trapped by his electric razor every day, that his beard grew much more rapidly each time he returned to the mainland and encountered girls. Schizophrenic patients are reported to have a special odor to their sweat, traced to trans-3-methylhexanoic acid.

The mind, already jelled by the advances in modern communication so that further boggling is impossible, twitches. One can imagine whole new industries springing up to create new perfumes ("A Scientific Combination of Primer and Releaser"), and other, larger corporations raising new turrets with flames alight at their tops on the Jersey flats, for

the production of phenolic, anesthetic, possibly bright
green sprays to cover, mask, or suppress all pheromones
("Don't Let On"). Gas chromatography of air samples
might reveal blips of difference between substances released
over a Glasgow football match, a committee meeting on
academic promotions, and a summer beach on Saturday
afternoon, all highly important. One can even imagine agitated conferences in the Pentagon, new agreements in Ge-
neva.

It is claimed that a well-trained tracking hound can follow
with accuracy the trail of a man in shoes, across open ground
marked by the footsteps of any number of other people,
provided the dog is given an item of the man's clothing to
smell beforehand. If one had to think up an R&D program
for a National Institute of Human Fragrance (to be created
by combining the budgets of the FDA and FCC), this would
be a good problem to start with. It might also provide the
kind of secondary, spin-off items of science that we like to
see in federally supported research. If it is true, as the novels
say, that an intelligent dog can tell the difference between
one human being and any other by detecting differences in
their scents, an explanation might be geometric differences
in 10-carbon molecules, or perhaps differences in the relative concentrations of several pheromones in a medley. If
this is a fact, it should be of interest to the immunologic
community, which has long since staked out claims on the
mechanisms involved in the discrimination between self and
non-self. Perhaps the fantastically sensitive and precise immunologic mechanisms for the detection of small molecules
such as haptenes represent another way of sensing the same
markers. Man's best friend might be used to sniff out his-
tocompatible donors. And so forth. If we could just succeed
in maintaining the research activity at this level, perhaps
diverting everyone's attention from all other aspects by
releasing large quantities of money, we might be able to stay
out of trouble.

The Music
of *This* Sphere

I t is one of our problems that as we become crowded
together, the sounds we make to each other, in our
increasingly complex communication systems, become
more random-sounding, accidental or incidental, and we
have trouble selecting meaningful signals out of the noise.
One reason is, of course, that we do not seem able to restrict
our communication to information-bearing, relevant signals. Given any new technology for transmitting information, we seem bound to use it for great quantities of small
talk. We are only saved by music from being overwhelmed
by nonsense.

It is a marginal comfort to know that the relatively new
science of bioacoustics must deal with similar problems in
the sounds made by other animals to each other. No matter
what sound-making device is placed at their disposal, creatures in general do a great deal of gabbling, and it requires
long patience and observation to edit out the parts lacking
syntax and sense. Light social conversation, designed to
keep the party going, prevails. Nature abhors a long silence.

Somewhere, underlying all the other signals, is a continual music. Termites make percussive sounds to each
other by beating their heads against the floor in the dark,
resonating corridors of their nests. The sound has been
described as resembling, to the human ear, sand falling on
paper, but spectrographic analysis of sound records has recently revealed a high degree of organization in the drum-

ming; the beats occur in regular, rhythmic phrases, differing in duration, like notes for a tympani section.

From time to time, certain termites make a convulsive movement of their mandibles to produce a loud, high-pitched clicking sound, audible ten meters off. So much effort goes into this one note that it must have urgent meaning, at least to the sender. He cannot make it without such a wrench that he is flung one or two centimeters into the air by the recoil.

There is obvious hazard in trying to assign a particular meaning to this special kind of sound, and problems like this exist throughout the field of bioacoustics. One can imagine a woolly-minded Visitor from Outer Space, interested in human beings, discerning on his spectrograph the click of that golf ball on the surface of the moon, and trying to account for it as a call of warning (unlikely), a signal of mating (out of the question), or an announcement of territory (could be).

Bats are obliged to make sounds almost ceaselessly, to sense, by sonar, all the objects in their surroundings. They can spot with accuracy, on the wing, small insects, and they will home onto things they like with infallibility and speed. With such a system for the equivalent of glancing around, they must live in a world of ultrasonic bat-sound, most of it with an industrial, machinery sound. Still, they communicate with each other as well, by clicks and high-pitched greetings. Moreover, they have been heard to produce, while hanging at rest upside down in the depths of woods, strange, solitary, and lovely bell-like notes.

Almost anything that an animal can employ to make a sound is put to use. Drumming, created by beating the feet, is used by prairie hens, rabbits, and mice; the head is banged by woodpeckers and certain other birds; the males of death-watch beetles make a rapid ticking sound by percussion of a protuberance on the abdomen against the ground; a faint but audible ticking is made by the tiny beetle *Lepinotus inquilinus,* which is less than two millimeters in length. Fish make sounds by clicking their teeth, blowing air, and drumming with special muscles against tuned inflated air bladders. Solid structures are set to vibrating by toothed bows in crustaceans and insects. The proboscis of the death's-head hawk moth is used as a kind of reed instrument, blown through to make high-pitched, reedy notes.

Gorillas beat their chests for certain kinds of discourse. Animals with loose skeletons rattle them, or, like rattlesnakes, get sounds from externally placed structures. Turtles, alligators, crocodiles, and even snakes make various more or less vocal sounds. Leeches have been heard to tap rhythmically on leaves, engaging the attention of other leeches, which tap back, in synchrony. Even earthworms make sounds, faint staccato notes in regular clusters. Toads sing to each other, and their friends sing back in antiphony.

Birdsong has been so much analyzed for its content of business communication that there seems little time left for music, but it is there. Behind the glossaries of warning calls, alarms, mating messages, pronouncements of territory, calls for recruitment, and demands for dispersal, there is redundant, elegant sound that is unaccountable as part of the working day. The thrush in my backyard sings down his nose in meditative, liquid runs of melody, over and over again, and I have the strongest impression that he does this for his own pleasure. Some of the time he seems to be practicing, like a virtuoso in his apartment. He starts a run, reaches a midpoint in the second bar where there should be a set of complex harmonics, stops, and goes back to begin over, dissatisfied. Sometimes he changes his notation so conspicuously that he seems to be improvising sets of variations. It is a meditative, questioning kind of music, and I cannot believe that he is simply saying, "thrush here."

The robin sings flexible songs, containing a variety of

motifs that he rearranges to his liking; the notes in each motif constitute the syntax, and the possibilities for variation produce a considerable repertoire. The meadow lark, with three hundred notes to work with, arranges these in phrases of three to six notes and elaborates fifty types of song. The nightingale has twenty-four basic songs, but gains wild variety by varying the internal arrangement of phrases and the length of pauses. The chaffinch listens to other chaffinches, and incorporates into his memory snatches of their songs.

The need to make music, and to listen to it, is universally expressed by human beings. I cannot imagine, even in our most primitive times, the emergence of talented painters to make cave paintings without there having been, near at hand, equally creative people making song. It is, like speech, a dominant aspect of human biology.

The individual parts played by other instrumentalists— crickets or earthworms, for instance—may not have the sound of music by themselves, but we hear them out of context. If we could listen to them all at once, fully orchestrated, in their immense ensemble, we might become aware of the counterpoint, the balance of tones and timbres and harmonics, the sonorities. The recorded songs of the humpback whale, filled with tensions and resolutions, ambiguities and allusions, incomplete, can be listened to as a *part* of music, like an isolated section of an orchestra. If we had better hearing, and could discern the descants of sea birds, the rhythmic tympani of schools of mollusks, or even the distant harmonics of midges hanging over meadows in the sun, the combined sound might lift us off our feet.

There are, of course, other ways to account for the songs of whales. They might be simple, down-to-earth statements about navigation, or sources of krill, or limits of territory. But the proof is not in, and until it is shown that these long, convoluted, insistent melodies, repeated by different singers with ornamentations of their own, are the means of sending through several hundred miles of undersea such ordinary information as "whale here," I shall believe otherwise. Now and again, in the intervals between songs, the whales have been seen to breach, leaping clear out of the sea and landing on their backs, awash in the turbulence of their beating flippers. Perhaps they are pleased by the way the piece went, or perhaps it is celebration at hearing one's own song returning after circumnavigation; whatever, it has the look of jubilation.

I suppose that my extraterrestrial Visitor night puzzle over my records in much the same way, on first listening. The 14th Quartet might, for him, be a communication announcing, "Beethoven here," answered, after passage through an undersea of time and submerged currents of human thought, by another long signal a century later, "Bartok here."

If, as I believe, the urge to make a kind of music is as much a characteristic of biology as our other fundamental functions, there ought to be an explanation for it. Having none at hand, I am free to make one up. The rhythmic sounds might be the recapitulation of something else—an earliest memory, a score for the transformation of inanimate, random matter in chaos into the improbable, ordered dance of living forms. Morowitz has presented the case, in thermodynamic terms, for the hypothesis that a steady flow of energy from the inexhaustible source of the sun to the unfillable sink of outer space, by way of the earth, is mathematically destined to cause the organization of matter into an increasingly ordered state. The resulting balancing act involves a ceaseless clustering of bonded atoms into molecules of higher and higher complexity, and the emergence of cycles for the storage and release of energy. In a nonequilibrium steady state, which is postulated, the solar energy would not just flow to the earth and radiate away; it is thermodynamically inevitable that it must rearrange matter into symmetry, away from probability, against entropy, lifting it, so to speak, into a constantly changing condition of rearrangement and molecular ornamentation. In such a system, the outcome is a chancy kind of order, always on the verge of descending into chaos, held taut against probability by the unremitting, constant surge of energy from the sun.

If there were to be sounds to represent this process, they would have the arrangement of the Brandenburg Concertos for my ear, but I am open to wonder whether the same events are recalled by the rhythms of insects, the long, pulsing runs of birdsong, the descants of whales, the modulated vibrations of a million locusts in migration, the tympani of gorilla breasts, termite heads, drumfish bladders. A "grand canonical ensemble" is, oddly enough, the proper term for a quantitative model system in thermodynamics, borrowed from music by way of mathematics. Borrowed back again, provided with notation, it would do for what I have in mind.

An Earnest Proposal

There was a quarter-page advertisement in the London *Observer* for a computer service that will enmesh your name in an electronic network of fifty thousand other names, sort out your tastes, preferences, habits, and deepest desires and match them up with opposite numbers, and retrieve for you, within a matter of seconds, and for a very small fee, friends. "Already," it says, "it [the computer] has given very real happiness and lasting relationships to thousands of people, and it can do the same for you!"

Without paying a fee, or filling out a questionnaire, all of us are being linked in similar circuits, for other reasons, by credit bureaus, the census, the tax people, the local police station, or the Army. Sooner or later, if it keeps on, the various networks will begin to touch, fuse, and then, in their coalescence, they will start sorting and retrieving each other, and we will all become bits of information on an enormous grid.

I do not worry much about the computers that are wired to help me find a friend among fifty thousand. If errors are made, I can always beg off with a headache. But what of the vaster machines that will be giving instructions to cities, to nations? If they are programmed to regulate human behavior according to today's view of nature, we are surely in for apocalypse.

The men who run the affairs of nations today are, by and

26

large, our practical men. They have been taught that the world is an arrangement of adversary systems, that force is what counts, aggression is what drives us at the core, only the fittest can survive, and only might can make more might. Thus, it is in observance of nature's law that we have planted, like perennial tubers, the numberless nameless missiles in the soil of Russia and China and our Midwestern farmlands, with more to come, poised to fly out at a nanosecond's notice, and meticulously engineered to ignite, in the centers of all our cities, artificial suns. If we let fly enough of them at once, we can even burn out the one-celled green creatures in the sea, and thus turn off the oxygen.

Before such things are done, one hopes that the computers will contain every least bit of relevant information about the way of the world. I should think we might assume this, in fairness to all. Even the nuclear realists, busy as their minds must be with calculations of acceptable levels of megadeath, would not want to overlook anything. They should be willing to wait, for a while anyway.

I have an earnest proposal to make. I suggest that we defer further action until we have acquired a really complete set of information concerning at least one living thing. Then, at least, we shall be able to claim that we know what we are doing. The delay might take a decade; let us say a decade. We and the other nations might set it as an objective of international, collaborative science to achieve a complete understanding of a single form of life. When this is done, and the information programmed into all our computers, I for one would be willing to take my chances.

As to the subject, I propose a simple one, easily solved within ten years. It is the protozoan *Myxotricha paradoxa*, which inhabits the inner reaches of the digestive tract of Australian termites.

It is not as though we would be starting from scratch. We have a fair amount of information about this creature al-

ready—not enough to understand him, of course, but
enough to inform us that he means something, perhaps a
great deal. At first glance, he appears to be an ordinary,
motile protozoan, remarkable chiefly for the speed and directness with which he swims from place to place, engulfing
fragments of wood finely chewed by his termite host. In the
termite ecosystem, an arrangement of Byzantine complexity, he stands at the epicenter. Without him, the wood,
however finely chewed, would never get digested; he sup-
plies the enzymes that break down cellulose to edible carbohydrate, leaving only the nondegradable lignin, which
the termite then excretes in geometrically tidy pellets and
uses as building blocks for the erection of arches and vaults
in the termite nest. Without him there would be no termites, no farms of the fungi that are cultivated by termites
and will grow nowhere else, and no conversion of dead
trees to loam.

The flagellae that beat in synchrony to propel myxotricha
with such directness turn out, on closer scrutiny with the
electron microscope, not to be flagellae at all. They are
outsiders, in to help with the business: fully formed, perfect
spirochetes that have attached themselves at regularly
spaced intervals all over the surface of the protozoan.

Then, there are oval organelles, embedded in the surface
close to the point of attachment of the spirochetes, and other
similar bodies drifting through the cytoplasm with the particles of still undigested wood. These, under high magnifica-
tion, turn out to be bacteria, living in symbiosis with the
spirochetes and the protozoan, probably contributing enzymes that break down the cellulose.

The whole animal, or ecosystem, stuck for the time being
halfway along in evolution, appears to be a model for the
development of cells like our own. Margulis has summa-
rized the now considerable body of data indicating that the
modern nucleated cell was made up, part by part, by the

coming together of just such prokaryotic animals. The blue-green algae, the original inventors of photosynthesis, entered partnership with primitive bacterial cells, and became the chloroplasts of plants; their descendants remain as discrete separate animals inside plant cells, with their own DNA and RNA, replicating on their own. Other bacteria with oxidative enzymes in their membranes, makers of ATP, joined up with fermenting bacteria and became the mitochondria of the future; they have since deleted some of their genes but retain personal genomes and can only be regarded as symbionts. Spirochetes, like the ones attached to *M. paradoxa,* joined up and became the cilia of eukaryotic cells. The centrioles, which hoist the microtubules on which chromosomes are strung for mitosis, are similar separate creatures; when not busy with mitosis, they become the basal bodies to which cilia are attached. And there are others, not yet clearly delineated, whose existence in the cell is indicated by the presence of cytoplasmic genes.

There is an underlying force that drives together the several creatures comprising myxotricha, and, then drives the assemblage into union with the termite. If we could understand this tendency, we would catch a glimpse of the process that brought single separate cells together for the construction of metazoans, culminating in the invention of roses, dolphins, and, of course, ourselves. It might turn out that the same tendency underlies the joining of organisms into communities, communities into ecosystems, and ecosystems into the biosphere. If this is, in fact, the drift of things, the way of the world, we may come to view immune reactions, genes for the chemical marking of self, and perhaps all reflexive responses of aggression and defense as secondary developments in evolution, necessary for the regulation and modulation of symbiosis, not designed to break into the process, only to keep it from getting out of hand.

If it is in the nature of living things to pool resources, to fuse when possible, we would have a new way of accounting for the progressive enrichment and complexity of form in living things.

I take it on faith that computers, although lacking souls, are possessed of a kind of intelligence. At the end of the decade, therefore, I am willing to predict that the feeding in of all the information then available will result, after a few seconds of whirring, in something like the following message, neatly and speedily printed out: "Request more data. How are spirochetes attached? Do not fire."

The Technology
of Medicine

Technology assessment has become a routine exercise for the scientific enterprises on which the country is obliged to spend vast sums for its needs. Brainy committees are continually evaluating the effectiveness and cost of doing various things in space, defense, energy, transportation, and the like, to give advice about prudent investments for the future.

Somehow medicine, for all the $80-odd billion that it is said to cost the nation, has not yet come in for much of this analytical treatment. It seems taken for granted that the technology of medicine simply exists, take it or leave it, and the only major technologic problem which policy-makers are interested in is how to deliver today's kind of health care, with equity, to all the people.

When, as is bound to happen sooner or later, the analysts get around to the technology of medicine itself, they will have to face the problem of measuring the relative cost and effectiveness of all the things that are done in the management of disease. They make their living at this kind of thing, and I wish them well, but I imagine they will have a bewildering time. For one thing, our methods of managing disease are constantly changing—partly under the influence of new bits of information brought in from all corners of biologic science. At the same time, a great many things are done that are not so closely related to science, some not related at all.

In fact, there are three quite different levels of technology in medicine, so unlike each other as to seem altogether different undertakings. Practitioners of medicine and the analysts will be in trouble if they are not kept separate.

1. First of all, there is a large body of what might be termed "nontechnology," impossible to measure in terms of its capacity to alter either the natural course of disease or its eventual outcome. A great deal of money is spent on this. It is valued highly by the professionals as well as the patients. It consists of what is sometimes called "supportive therapy." It tides patients over through diseases that are not, by and large, understood. It is what is meant by the phrases "caring for" and "standing by." It is indispensable. It is not, however, a technology in any real sense, since it does not involve measures directed at the underlying mechanism of disease.

It includes the large part of any good doctor's time that is taken up with simply providing reassurance, explaining to patients who fear that they have contracted one or another lethal disease that they are, in fact, quite healthy.

It is what physicians used to be engaged in at the bedside of patients with diphtheria, meningitis, poliomyelitis, lobar pneumonia, and all the rest of the infectious diseases that have since come under control.

It is what physicians must now do for patients with intractable cancer, severe rheumatoid arthritis, multiple sclerosis, stroke, and advanced cirrhosis. One can think of at least twenty major diseases that require this kind of supportive medical care because of the absence of an effective technology. I would include a large amount of what is called mental disease, and most varieties of cancer, in this category.

The cost of this nontechnology is very high, and getting higher all the time. It requires not only a great deal of time but also very hard effort and skill on the part of physicians;

only the very best of doctors are good at coping with this kind of defeat. It also involves long periods of hospitalization, lots of nursing, lots of involvement of nonmedical professionals in and out of the hospital. It represents, in short, a substantial segment of today's expenditures for health.

2. At the next level up is a kind of technology best termed "halfway technology." This represents the kinds of things that must be done after the fact, in efforts to compensate for the incapacitating effects of certain diseases whose course one is unable to do very much about. It is a technology designed to make up for disease, or to postpone death.

The outstanding examples in recent years are the transplantations of hearts, kidneys, livers, and other organs, and the equally spectacular inventions of artificial organs. In the public mind, this kind of technology has come to seem like the equivalent of the high technologies of the physical sciences. The media tend to present each new procedure as though it represented a breakthrough and therapeutic triumph, instead of the makeshift that it really is.

In fact, this level of technology is, by its nature, at the same time highly sophisticated and profoundly primitive. It is the kind of thing that one must continue to do until there is a genuine understanding of the mechanisms involved in disease. In chronic glomerulonephritis, for example, a much clearer insight will be needed into the events leading to the destruction of glomeruli by the immunologic reactants that now appear to govern this disease, before one will know how to intervene intelligently to prevent the process, or turn it round. But when this level of understanding has been reached, the technology of kidney replacement will not be much needed and should no longer pose the huge problems of logistics, cost, and ethics that it poses today.

An extremely complex and costly technology for the management of coronary heart disease has evolved—involving specialized ambulances and hospital units, all kinds of electronic gadgetry, and whole platoons of new professional personnel—to deal with the end results of coronary thrombosis. Almost everything offered today for the treatment of heart disease is at this level of technology, with the transplanted and artificial hearts as ultimate examples. When enough has been learned to know what really goes wrong in heart disease, one ought to be in a position to figure out ways to prevent or reverse the process, and when this happens the current elaborate technology will probably be set to one side.

Much of what is done in the treatment of cancer, by surgery, irradiation, and chemotherapy, represents halfway technology, in the sense that these measures are directed at the existence of already established cancer cells, but not at the mechanisms by which cells become neoplastic.

It is a characteristic of this kind of technology that it costs an enormous amount of money and requires a continuing expansion of hospital facilities. There is no end to the need for new, highly trained people to run the enterprise. And there is really no way out of this, at the present state of knowledge. If the installation of specialized coronary-care units can result in the extension of life for only a few patients with coronary disease (and there is no question that this technology is effective in a few cases), it seems to me an inevitable fact of life that as many of these as can be will be put together, and as much money as can be found will be spent. I do not see that anyone has much choice in this. The only thing that can move medicine away from this level of technology is new information, and the only imaginable source of this information is research.

3. The third type of technology is the kind that is so effective that it seems to attract the least public notice; it has come to be taken for granted. This is the genuinely decisive technology of modern medicine, exemplified best by mod-

ern methods for immunization against diphtheria, pertussis, and the childhood virus diseases, and the contemporary use of antibiotics and chemotherapy for bacterial infections. The capacity to deal effectively with syphilis and tuberculosis represents a milestone in human endeavor, even though full use of this potential has not yet been made. And there are, of course, other examples: the treatment of endocrinologic disorders with appropriate hormones, the prevention of hemolytic disease of the newborn, the treatment and prevention of various nutritional disorders, and perhaps just around the corner the management of Parkinsonism and sickle-cell anemia. There are other examples, and everyone will have his favorite candidates for the list, but the truth is that there are nothing like as many as the public has been led to believe.

The point to be made about this kind of technology—the real high technology of medicine—is that it comes as the result of a genuine understanding of disease mechanisms, and when it becomes available, it is relatively inexpensive, relatively simple, and relatively easy to deliver.

Offhand, I cannot think of any important human disease for which medicine possesses the outright capacity to prevent or cure where the cost of the technology is itself a major problem. The price is never as high as the cost of managing the same diseases during the earlier stages of no-technology or halfway technology. If a case of typhoid fever had to be managed today by the best methods of 1935, it would run to a staggering expense. At, say, around fifty days of hospitalization, requiring the most demanding kind of nursing care, with the obsessive concern for details of diet that characterized the therapy of that time, with daily laboratory monitoring, and, on occasion, surgical intervention for abdominal catastrophe, I should think $10,000 would be a conservative estimate for the illness, as contrasted with today's cost of a bottle of chloramphenicol and a day or two of fever. The halfway technology that was evolving for poliomyelitis in the early 1950s, just before the emergence of the basic research that made the vaccine possible, provides another illustration of the point. Do you remember Sister Kenny, and the cost of those institutes for rehabilitation, with all those ceremonially applied hot fomentations, and the debates about whether the affected limbs should be totally immobilized or kept in passive motion as frequently as possible, and the masses of statistically tormented data mobilized to support one view or the other? It is the cost of that kind of technology, and its relative effectiveness, that must be compared with the cost and effectiveness of the vaccine.

Pulmonary tuberculosis had similar episodes in its history. There was a sudden enthusiasm for the surgical removal of infected lung tissue in the early 1950s, and elaborate plans were being made for new and expensive installations for major pulmonary surgery in tuberculosis hospitals, and then INH and streptomycin came along and the hospitals themselves were closed up.

It is when physicians are bogged down by their incomplete technologies, by the innumerable things they are obliged to do in medicine when they lack a clear understanding of disease mechanisms, that the deficiencies of the health-care system are most conspicuous. If I were a policymaker, interested in saving money for health care over the long haul, I would regard it as an act of high prudence to give high priority to a lot more basic research in biologic science. This is the only way to get the full mileage that biology owes to the science of medicine, even though it seems, as used to be said in the days when the phrase still had some meaning, like asking for the moon.

Vibes

We leave traces of ourselves wherever we go, on whatever we touch. One of the odd discoveries made by small boys is that when two pebbles are struck sharply against each other they emit, briefly, a curious smoky odor. The phenomenon fades when the stones are immaculately cleaned, vanishes when they are heated to furnace temperature, and reappears when they are simply touched by the hand again before being struck.

An intelligent dog with a good nose can track a man across open ground by his smell and distinguish that man's tracks from those of others. More than this, the dog can detect the odor of a light human fingerprint on a glass slide, and he will remember that slide and smell it out from others for as long as six weeks, when the scent fades away. Moreover, this animal can smell the identity of identical twins, and will follow the tracks of one or the other as though they had been made by the same man.

We are marked as self by the chemicals we leave beneath the soles of our shoes, as unmistakably and individually as by the membrane surface antigens detectable in homografts of our tissues.

Other animals are similarly endowed with signaling mechanisms. Columns of ants can smell out the differences between themselves and other ants on their trails. The ants of one species, proceeding jerkily across a path, leave trails that can be followed by their own relatives but not by others. Certain ants, predators, have taken unfair advantage of the system; they are born with an ability to sense the trails of the species they habitually take for slaves, follow their victims to their nests, and release special odorants that throw them into disorganized panic.

Minnows and catfish can recognize each member of their own species by his particular, person-specific odor. It is hard to imagine a solitary, independent, existentialist minnow, recognizable for himself alone; minnows in a school behave like interchangeable, identical parts of an organism. But there it is.

The problem of olfactory sensing shares some of the current puzzles and confusions of immunology, apart from the business of telling self from non-self. A rabbit, it has been calculated, has something like 100 million olfactory receptors. There is a constant and surprisingly rapid turnover of the receptor cells, with new ones emerging from basal cells within a few days. The theories to explain olfaction are as numerous and complex as those for immunologic sensing. It seems likely that the shape of the smelled molecule is what matters most. By and large, odorants are chemically small, Spartan compounds. In a rose garden, a rose is a rose because of geraniol, a 10-carbon compound, and it is the geometric conformation of atoms and their bond angles that determine the unique fragrance. The special vibrations of atoms or groups of atoms within the molecules of odorants, or the vibratory song of the entire molecule, have been made the basis for several theories, with postulated "osmic frequencies" as the source of odor. The geometry of the molecule seems to be more important than the names of the atoms themselves; any set of atoms, if arranged in precisely the same configuration, by whatever chemical name, might smell as sweet. It is not known how the olfactory cells are fired by an odorant. According to one view, a hole is poked in the receptor membrane, launching depolarization, but other workers believe that the substance may become bound to the cells possessing specific receptors for it and then may just sit there, somehow displaying its signal from a distance, after the fashion of antigens on immune cells. Specific receptor proteins have been proposed, with different olfactory cells carrying specific receptors for different "primary" odors, but no one has yet succeeded in identifying the receptors or naming the "primary" odors.

Training of cells for olfactory sensing appears to be an everyday phenomenon. Repeated exposure of an animal to the same odorant, in small doses, leads to great enhancement of acuity, suggesting the possibility that new receptor sites are added to the cells. It is conceivable that new clones of cells with a particular receptor are stimulated to emerge in the process of training. The guinea pig, that immunologically famous animal, can be trained to perceive fantastically small amounts of nitrobenzene by his nose, without the help of Freund's adjuvants or haptene carriers. Minnows have been trained to recognize phenol, and distinguish it from p-chlorophenol, in concentrations of five parts per billion. Eels have been taught to smell two or three molecules of phenylethyl alcohol. And, of course, eels and salmon must be able to remember by nature, as the phrase goes, the odor of the waters in which they were hatched, so as to sniff their way back from the open sea for spawning. Electrodes in the olfactory bulbs of salmon will fire when the olfactory epithelium is exposed to water from their spawning grounds, whereas water from other streams causes no response.

We feel somehow inferior and left out of things by all the marvelous sensory technology in the creatures around us. We sometimes try to diminish our sense of loss (or loss of sense) by claiming to ourselves that we have put such primitive mechanisms behind us in our evolution. We like to regard the olfactory bulb as a sort of archeologic find, and we speak of the ancient olfactory parts of the brain as though they were elderly, dotty relatives in need of hobbies.

But we may be better at it than we think. An average man

can detect just a few molecules of butyl mercaptan, and most of us can sense the presence of musk in vanishingly small amounts. Steroids are marvelously odorous, emitting varieties of musky, sexy smells. Women are acutely aware of the odor of a synthetic steroid named exaltolide, which most men are unable to detect. All of us are able to smell ants, for which the great word pismire was originally coined.

There may even be odorants that fire off receptors in our olfactory epithelia without our being conscious of smell, including signals exchanged involuntarily between human beings. Wiener has proposed, on intuitive grounds, that defects and misinterpretations in such a communication system may be an unexplored territory for psychiatry. The schizophrenic, he suggests, may have his problems with identity and reality because of flawed perceptions of his own or others' signals. And, indeed, there may be something wrong with the apparatus in schizophrenics; they have, it is said, an unfamiliar odor, recently attributed to trans-3-methylhexanoic acid, in their sweat.

Olfactory receptors for communication between different creatures are crucial for the establishment of symbiotic relations. The crab and anemone recognize each other as partners by molecular configurations, as do the anemones and their symbiotic damsel fish. Similar devices are employed for defense, as with the limpet, which defends itself against starfish predators by everting its mantle and thus precluding a starfish foothold; the limpet senses a special starfish protein, which is, perhaps in the name of fairness, elaborated by all starfish into their environment. The system is evidently an ancient one, long antedating the immunologic sensing of familiar or foreign forms of life by the antibodies on which we now depend so heavily for our separateness. It has recently been learned that the genes for the marking of self by cellular antigens and those for making immunologic responses by antibody formation are closely linked. It is possible that the invention of antibodies evolved from the earlier sensing mechanisms needed for symbiosis, perhaps designed, in part, to keep the latter from getting out of hand.

A very general system of chemical communication between living things of all kinds, plant and animal, has been termed "allelochemics" by Whittaker. Using one signal or another, each form of life announces its proximity to the others around it, setting limits on encroachment or spreading welcome to potential symbionts. The net effect is a coordinated mechanism for the regulation of rates of growth and occupations of territory. It is evidently designed for the homeostasis of the earth.

Jorge Borges, in his recent bestiary of mythical creatures, notes that the idea of round beasts was imagined by many speculative minds, and Johannes Kepler once argued that the earth itself is such a being. In this immense organism, chemical signals might serve the function of global hormones, keeping balance and symmetry in the operation of various interrelated working parts, informing tissues in the vegetation of the Alps about the state of eels in the Sargasso Sea, by long, interminable relays of interconnected messages between all kinds of other creatures.

This is an interesting kind of problem, made to order for computers if they came in sizes big enough to store in nearby galaxies. It is nice to think that there are so many unsolved puzzles ahead for biology, although I wonder whether we will ever find enough graduate students.

Ceti

Tau Ceti is a relatively nearby star that sufficiently resembles our sun to make its solar system a plausible candidate for the existence of life. We are, it appears, ready to begin getting in touch with Ceti, and with any other interested celestial body in more remote places, out to the edge. CETI is also, by intention, the acronym of the First International Conference on Communication with Extraterrestrial Intelligence, held in 1972 in Soviet Armenia under the joint sponsorship of the National Academy of Sciences of the United States and the Soviet Academy which involved eminent physicists and astronomers from various countries, most of whom are convinced that the odds for the existence of life elsewhere are very high, with a reasonable probability that there are civilizations, one place or another, with technologic mastery matching or exceeding ours.

On this assumption, the conferees thought it likely that radioastronomy would be the generally accepted mode of interstellar communication, on grounds of speed and economy. They made a formal recommendation that we organize an international cooperative program, with new and immense radio telescopes, to probe the reaches of deep space for electromagnetic signals making sense. Eventually, we would plan to send out messages on our own and receive answers, but at the outset it seems more practical to begin by catching snatches of conversation between others.

So, the highest of all our complex technologies in the hardest of our sciences will soon be engaged, full scale, in what is essentially biologic research—and with some aspects of social science, at that.

The earth has become, just in the last decade, too small a place. We have the feeling of being confined—shut in; it is something like outgrowing a small town in a small county. The views of the dark, pocked surface of Mars, still lifeless to judge from the latest photographs, do not seem to have extended our reach; instead, they bring closer, too close, another unsatisfactory feature of our local environment. The blue noonday sky, cloudless, has lost its old look of immensity. The word is out that the sky is not limitless; it is finite. It is, in truth, only a kind of local roof, a membrane under which we live, luminous but confusingly refractile when suffused with sunlight; we can sense its concave surface a few miles over our heads. We know that it is tough and thick enough so that when hard objects strike it from the outside they burst into flames. The color photographs of the earth are more amazing than anything outside: we live inside a blue chamber, a bubble of air blown by ourselves. The other sky beyond, absolutely black and appalling, is wide-open country, irresistible for exploration.

Here we go, then. An extraterrestrial embryologist, having a close look at us from time to time, would probably conclude that the morphogenesis of the earth is coming along well, with the beginnings of a nervous system and fair-sized ganglions in the form of cities, and now with specialized, dish-shaped sensory organs, miles across, ready to receive stimuli. He may well wonder, however, how we will go about responding. We are evolving into the situation of a Skinner pigeon in a Skinner box, peering about in all directions, trying to make connections, probing.

When the first word comes in from outer space, finally, we will probably be used to the idea. We can already provide a quite good explanation for the origin of life, here or elsewhere. Given a moist planet with methane, formaldehyde, ammonia, and some usable minerals, all of which abound, exposed to lightning or ultraviolet irradiation at the right temperature, life might start off almost anywhere. The tricky, unsolved thing is how to get the polymers to arrange in membranes and invent replication. The rest is clear going. If they follow our protocol, it will be anaerobic life at first, then photosynthesis and the first exhalation of oxygen, then respiring life and the great burst of variation, then speciation, and, finally, some kind of consciousness. It is easy, in the telling.

I suspect that when we have recovered from the first easy acceptance of signs of life from elsewhere, and finished nodding at each other, and finished smiling, we will be in for shock. We have had it our way, relatively speaking, being unique all these years, and it will be hard to deal with the thought that the whole, infinitely huge, spinning, clock-like apparatus around us is itself animate, and can sprout life whenever the conditions are right. We will respond, beyond doubt, by making connections after the fashion of established life, floating out our filaments, extending pili, but we will end up feeling smaller than ever, as small as a single cell, with a quite new sense of continuity. It will take some getting used to.

The immediate problem, however, is a much more practical, down-to-earth matter, and must be giving insomnia to the CETI participants. Let us assume that there is, indeed, sentient life in one or another part of remote space, and that we will be successful in getting in touch with it. What on earth are we going to talk about? If, as seems likely, it is a hundred or more light years away, there are going to be some very long pauses. The barest amenities, on which we rely for opening conversations—Hello, are you there?, from us, followed by Yes, hello, from them—will take two hundred years at least. By the time we have our party we may have forgotten what we had in mind.

We could begin by gambling on the rightness of our

technology and just send out news of ourselves, like a mimeographed Christmas letter, but we would have to choose our items carefully, with durability of meaning in mind. Whatever information we provide must still make sense to us two centuries later, and must still seem important, or the conversation will be an embarrassment to all concerned. In two hundred years it is, as we have found, easy to lose the thread.

Perhaps the safest thing to do at the outset, if technology permits, is to send music. This language may be the best we have for explaining what we are like to others in space, with least ambiguity. I would vote for Bach, all of Bach, streamed out into space, over and over again. We would be bragging, of course, but it is surely excusable for us to put the best possible face on at the beginning of such an acquaintance. We can tell the harder truths later. And, to do ourselves justice, music would give a fairer picture of what we are really like than some of the other things we might be sending, like *Time,* say, or a history of the U.N. or Presidential speeches. We could send out our science, of course, but just think of the wincing at this end when the polite comments arrive two hundred years from now. Whatever we offer as today's items of liveliest interest are bound to be out of date and irrelevant, maybe even ridiculous. I think we should stick to music.

Perhaps, if the technology can be adapted to it, we should send some paintings. Nothing would better describe what this place is like, to an outsider, than the Cézanne demonstrations that an apple is really part fruit, part earth.

What kinds of questions should we ask? The choices will be hard, and everyone will want his special question first. What are your smallest particles? Did you think yourselves unique? Do you have colds? Have you anything quicker than light? Do you always tell the truth? Do you cry? There is no end to the list.

Perhaps we should wait a while, until we are sure we

know what we want to know, before we get down to detailed questions. After all, the main question will be the opener: Hello, are you there? If the reply should turn out to be Yes, hello, we might want to stop there and think about that, for quite a long time.

The Long
Habit

We continue to share with our remotest ancestors the most tangled and evasive attitudes about death, despite the great distance we have come in understanding some of the profound aspects of biology. We have as much distaste for talking about personal death as for thinking about it; it is an indelicacy, like talking in mixed company about venereal disease or abortion in the old days. Death on a grand scale does not bother us in the same special way: we can sit around a dinner table and discuss war, involving 60 million volatilized human deaths, as though we were talking about bad weather; we can watch abrupt bloody death every day, in color, on films and television, without blinking back a tear. It is when the numbers of dead are very small, and very close, that we begin to think in scurrying circles. At the very center of the problem is the naked cold deadness of one's own self, the only reality in nature of which we can have absolute certainty, and it is unmentionable, unthinkable. We may be even less willing to face the issue at first hand than our predecessors because of a secret new hope that maybe it will go away. We like to think, hiding the thought, that with all the marvelous ways in which we seem now to lead nature around by the nose, perhaps we can avoid the central problem if we just become, next year, say, a bit smarter.

"The long habit of living," said Thomas Browne, "indisposeth us to dying." These days, the habit has become an addiction: we are hooked on living; the tenacity of its grip on us, and ours on it, grows in intensity. We cannot think of giving it up, even when living loses its zest—even when we have lost the zest for zest.

We have come a long way in our technologic capacity to put death off, and it is imaginable that we might learn to stall it for even longer periods, perhaps matching the life-spans of the Abkhasian Russians, who are said to go on, springily, for a century and a half. If we can rid ourselves of some of our chronic, degenerative diseases, and cancer, strokes, and coronaries, we might go on and on. It sounds attractive and reasonable, but it is no certainty. If we became free of disease, we would make a much better run of it for the last decade or so, but might still terminate on about the same schedule as now. We may be like the genetically different lines of mice, or like Hayflick's different tissue-culture lines, programmed to die after a predetermined number of days, clocked by their genomes. If this is the way it is, some of us will continue to wear out and come unhinged in the sixth decade, and some much later, depending on genetic time-tables.

If we ever do achieve freedom from most of today's diseases, or even complete freedom from disease, we will perhaps terminate by drying out and blowing away on a light breeze, but we will still die.

Most of my friends do not like this way of looking at it. They prefer to take it for granted that we only die because we get sick, with one lethal ailment or another, and if we did not have our diseases we might go on indefinitely. Even biologists choose to think this about themselves, despite the evidences of the absolute inevitability of death that surround their professional lives. Everything dies, all around, trees, plankton, lichens, mice, whales, flies, mitochondria. In the simplest creatures it is sometimes difficult to see it as death, since the strands of replicating DNA they leave behind are more conspicuously the living parts of themselves than with us (not that it is fundamentally any different, but it seems so). Flies do not develop a ward round of diseases that carry them off, one by one. They simply age, and die, like flies.

We hanker to go on, even in the face of plain evidence that long, long lives are not necessarily pleasurable in the kind of society we have arranged thus far. We will be lucky if we can postpone the search for new technologies for a while, until we have discovered some satisfactory things to do with the extra time. Something will surely have to be found to take the place of sitting on the porch re-examining one's watch.

Perhaps we would not be so anxious to prolong life if we did not detest so much the sickness of withdrawal. It is astonishing how little information we have about this universal process, with all the other dazzling advances in biology. It is almost as though we wanted not to know about it. Even if we could imagine the act of death in isolation, without any preliminary stage of being struck down by disease, we would be fearful of it.

There are signs that medicine may be taking a new interest in the process, partly from curiosity, partly from an embarrassed realization that we have not been handling this aspect of disease with as much skill as physicians once displayed, back in the days before they became convinced that disease was their solitary and sometimes defeatable enemy. It used to be the hardest and most important of all the services of a good doctor to be on hand at the time of death and to provide comfort, usually in the home. Now it is done in hospitals, in secrecy (one of the reasons for the increased fear of death these days may be that so many people are totally unfamiliar with it; they never actually see it happen in real life). Some of our technology permits us to deny its existence, and we maintain flickers of life for long stretches in one community of cells or another, as though we were keeping a flag flying. Death is not a sudden-all-at-once affair; cells go down in sequence, one by one. You can, if you like, recover great numbers of them many hours after the lights have gone out, and grow them out in cultures. It takes hours, even days, before the irreversible word finally gets around to all the provinces.

We may be about to rediscover that dying is not such a bad thing to do after all. Sir William Osler took this view: he disapproved of people who spoke of the agony of death, maintaining that there was no such thing.

In a nineteenth-century memoir on an expedition in Africa, there is a story by David Livingston about his own experience of near-death. He was caught by a lion, crushed across the chest in the animal's great jaws, and saved in the instant by a lucky shot from a friend. Later, he remembered the episode in clear detail. He was so amazed by the extraordinary sense of peace, calm, and total painlessness associated with being killed that he constructed a theory that all creatures are provided with a protective physiologic mechanism, switched on at the verge of death, carrying them through in a haze of tranquillity.

I have seen agony in death only once, in a patient with rabies; he remained acutely aware of every stage in the process of his own disintegration over a twenty-four-hour period, right up to his final moment. It was as though, in the special neuropathology of rabies, the switch had been prevented from turning.

We will be having new opportunities to learn more about the physiology of death at first hand, from the increasing numbers of cardiac patients who have been through the whole process and then back again. Judging from what has been found out thus far, from the first generation of people resuscitated from cardiac standstill (already termed the Lazarus syndrome), Osler seems to have been right. Those who remember parts or all of their episodes do not recall any fear, or anguish. Several people who remained conscious throughout, while appearing to have been quite dead, could only describe a remarkable sensation of detachment. One man underwent coronary occlusion with cessation of the heart and dropped for all practical purposes dead, in front of a hospital; within a few minutes his heart had been restarted by electrodes and he breathed his way back into life. According to his account, the strangest thing was that there were so many people around him, moving so urgently, handling his body with such excitement, while all his awareness was of quietude.

In a recent study of the reaction to dying in patients with obstructive disease of the lungs, it was concluded that the process was considerably more shattering for the professional observers than the observed. Most of the patients appeared to be preparing themselves with equanimity for death, as though intuitively familiar with the business. One elderly woman reported that the only painful and distressing part of the process was in being interrupted; on several occasions she was provided with conventional therapeutic measures to maintain oxygenation or restore fluids and electrolytes, and each time she found the experience of coming back harrowing; she deeply resented the interference with her dying.

I find myself surprised by the thought that dying is an all-right thing to do, but perhaps it should not surprise. It is, after all, the most ancient and fundamental of biologic functions, with its mechanisms worked out with the same attention to detail, the same provision for the advantage of the organism, the same abundance of genetic information for guidance through the stages, that we have long since become accustomed to finding in all the crucial acts of living.

Very well. But even so, if the transformation is a coordinated, integrated physiologic process in its initial, local stages, there is still that permanent vanishing of consciousness to be accounted for. Are we to be stuck forever with this problem? Where on earth does it go? Is it simply stopped dead in its tracks, lost in humus, wasted? Considering the tendency of nature to find uses for complex and intricate mechanisms, this seems to me unnatural. I prefer to think of it as somehow separated off at the filaments of its attachment, and then drawn like an easy breath back into the membrane of its origin, a fresh memory for a biospherical nervous system, but I have no data on the matter.

This is for another science, another day. It may turn out, as some scientists suggest, that we are forever precluded from investigating consciousness by a sort of indeterminacy principle that stipulates that the very act of looking will make it twitch and blur out of sight. It this is true, we will never learn. I envy some of my friends who are convinced about telepathy; oddly enough, it is my European scientist acquaintances who believe it most freely and take it most lightly. All their aunts have received Communications, and there they sit, with proof of the motility of consciousness at their fingertips, and the making of a new science. It is discouraging to have had the wrong aunts, and never the ghost of a message.

Antaeus in
Manhattan

I nsects again.

When social animals are gathered together in groups, they become qualitatively different creatures from what they were when alone or in pairs. Single locusts are quiet, meditative, sessile things, but when locusts are added to other locusts, they become excited, change color, undergo spectacular endocrine revisions, and intensify their activity until, when there are enough of them packed shoulder to shoulder, they vibrate and hum with the energy of a jet airliner and take off.

Watson, Nel, and Hewitt have collected large numbers of termites in the field and placed them together for observation, in groups and pairs. The grouped termites become increasingly friendly and active, but show no inclination to lay eggs or mate; instead, they cut down on their water intake, watching their weight, and the mitochondria of their flight muscles escalate in metabolic activity. Grouped termites keep touching each other incessantly with their antennae, and this appears to be the central governing mechanism. It is the being touched that counts, rather than the act of touching. Deprived of antennae, any termite can become a group termite if touched frequently enough by the others.

Isolated, paired termites are something else again. As soon as they are removed from the group, and the touching from all sides comes to an end, they become aggressive, standoffish; they begin drinking compulsively, and abstain from touching each other. Sometimes, they even bite off the distal halves of each other's antennae, to eliminate the temptation. Irritably, settling down to make the best of a poor situation, they begin preparations for the laying of eggs and the taking care of the brood. Meanwhile, the mitochondria in their flight muscles go out of business.

The most intensely social animals can only adapt to group behavior. Bees and ants have no option when isolated, except to die. There is really no such creature as a single individual; he has no more life of his own than a cast-off cell marooned from the surface of your skin.

Ants are more like the parts of an animal than entities on their own. They are mobile cells, circulating through a dense connective tissue of other ants in a matrix of twigs. The circuits are so intimately woven that the anthill meets all the essential criteria of an organism.

It would be wonderful to understand how the anthill communication system works. Somehow, by touching each other continually, by exchanging bits of white stuff carried about in their mandibles like money, they manage to inform the whole enterprise about the state of the world outside, the location of food, the nearness of enemies, the maintenance requirements of the Hill, even the direction of the sun; in the Alps, mountaineers are said to use the ameboid configurations of elongated ant nests as pointers to the south. The Hill, for its part, responds by administering the affairs of the institution, coordinating and synchronizing the movements of its crawling parts, aerating and cleaning the nest so that it can last for as long as forty years, fetching food in by long tentacles, rearing broods, taking slaves, raising crops, and, at one time or another, budding off subcolonies in the near vicinity, as progeny.

The social insects, especially ants, have been sources of all kinds of parables, giving lessons in industry, interdepend-

ence, altruism, humility, frugality, patience. They have been
employed to instruct us in the whole range of our institutional virtues, from the White House to your neighborhood
savings bank.

And now, at last, they have become an Art Form. A
gallery in New York exhibited a collection of 2 million live
army ants, on loan from Central America, in a one-colony
show entitled "Patterns and Structures." They were displayed on sand in a huge square bin, walled by plastic sides
high enough to prevent them from crawling over and out
into Manhattan. The inventor of the work arranged and
rearranged the location of food sources in different places,
according to his inspiration and their taste, and they formed
themselves into long, black, ropy patterns, extended like
writhing limbs, hands, fingers, across the sand in crescents,
crisscrosses, and long ellipses, from one station to another.
Thus deployed, they were watched with intensity by the
crowds of winter-carapaced people who lined up in neat
rows to gaze down at them. The ants were, together with
the New Yorkers, an abstraction, a live mobile, an action
painting, a piece of found art, a happening, a parody, depending on the light.

I can imagine the people moving around the edges of the
plastic barrier, touching shoulder to shoulder, sometimes
touching hands, exchanging bits of information, nodding,
smiling sometimes, prepared as New Yorkers always are to
take flight at a moment's notice, their mitochondria fully
stoked and steaming. They move in orderly lines around the
box, crowding one another precisely, without injury, peering down, nodding, and then backing off to let new people
in. Seen from a distance, clustered densely around the white
plastic box containing the long serpentine lines of army ants,
turning to each other and murmuring repetitively, they
seem an absolute marvel. They might have dropped here
from another planet.

I am sad that I did not see any of this myself. By the time I had received the communication on television and in my morning paper, felt the tugging pull toward Manhattan, and made my preparations to migrate, I learned that the army ants had all died.

The Art Form simply disintegrated, all at once, like one of those exploding, vanishing faces in paintings by the British artist Francis Bacon.

There was no explanation, beyond the rumored, unproved possibility of cold drafts in the gallery over the weekend. Monday morning they were sluggish, moving with less precision, dully. Then, the death began, affecting first one part and then another, and within a day all 2 million were dead, swept away into large plastic bags and put outside for engulfment and digestion by the sanitation truck.

It is a melancholy parable. I am unsure of the meaning, but I do think it has something to do with all that plastic— that, and the distance from the earth. It is a long, long way from the earth of a Central American jungle to the ground floor of a gallery, especially when you consider that Manhattan itself is suspended on a kind of concrete platform, propped up by a meshwork of wires, pipes, and water mains. But I think it was chiefly the plastic, which seems to me the most unearthly of all man's creations so far. I do not believe you can suspend army ants away from the earth, on plastic, for any length of time. They will lose touch, run out of energy, and die for lack of current.

One steps on ants, single ants or small clusters, every day without giving it a thought, but it is impossible to contemplate the death of so vast a beast as these 2 million ants without feeling twinges of sympathy, and something else. Nervously, thinking this way, thinking especially about Manhattan and the plastic platform, I laid down my newspaper and reached for the book on my shelf that contained, I knew, precisely the paragraph of reassurance required by the moment:

It is not surprising that many analogies have been drawn between the social insects and human societies. Fundamentally, however, these are misleading or meaningless, for the behavior of insects is rigidly stereotyped and determined by innate instructive mechanisms; they show little or no insight or capacity for learning, and they lack the ability to develop a social tradition based on the accumulated experience of many generations.

It is, of course, an incomplete comfort to read this sort of thing to one's self. For full effect, it needs reading aloud by several people at once, moving the lips in synchrony.

The MBL

O nce you have become permanently startled, as I am, by the realization that we are a social species, you tend to keep an eye out for pieces of evidence that this is, by and large, a good thing for us. You look around for the enterprises that we engage in collectively and unconsciously, the things we build like wasp nests, individually unaware of what we are doing. Most of the time, these days, it is a depressing exercise. The joint building activity that consumes most of our energy and binds us together is, of course, language, but this is so overwhelming a structure and grows so slowly that none of us can feel a personal sense of participating in the work.

The less immense, more finite items, of a size allowing the mind to get a handhold, like nations, or space technology, or New York, are hard to think about without drifting toward heartsink.

It is in our very small enterprises that we can find encouragement, here and there. The Marine Biological Laboratory in Woods Hole is a paradigm, a human institution possessed of a life of its own, self-regenerating, touched all around by human meddle but constantly improved, embellished by it. The place was put together, given life, sustained into today's version of its maturity and prepared for further elaboration and changes in its complexity, by what can only be described as a bunch of people. Neither the spectacularly eminent men who have served as directors down through the century nor the numberless committees by which it is seasonally raddled, nor the six hundred-man corporation that nominally owns and operates it, nor even the trustees, have ever been able to do more than hold the lightest reins over this institution; it seems to have a mind of its own, which it makes up in its own way.

Successive generations of people in bunches, never seeming very well organized, have been building the MBL since it was chartered in 1888. It actually started earlier, in 1871, when Woods Hole, Massachusetts, was selected for a Bureau of Fisheries Station and the news got round that all sorts of marine and estuarine life could be found here in the collisions between the Gulf Stream and northern currents offshore, plus birds to watch. Academic types drifted down from Boston, looked around, began explaining things to each other, and the place was off and running.

The MBL has grown slowly but steadily from the outset, sprouting new buildings from time to time, taking on new functions, expanding, drawing to itself by a sort of tropism greater numbers of biological scientists each summer, attracting students from all parts of the world. Today, it stands as the uniquely national center for biology in this country; it is the National Biological Laboratory without being officially designated (or yet funded) as such. Its influence on the growth and development of biologic science has been equivalent to that of many of the country's universities combined, for it has had its pick of the world's scientific talent for each summer's research and teaching. If you ask around, you will find that any number of today's leading figures in biology and medicine were informally ushered into their careers by the summer course in physiology; a still greater number picked up this or that idea for their key experiments while spending time as summer visitors in the laboratories, and others simply came for a holiday and got enough good notions to keep their laboratories back home busy for a full year. Someone has counted thirty Nobel Laureates who have worked at the MBL at one time or another.

It is amazing that such an institution, exerting so much influence on academic science, has been able to remain so absolutely autonomous. It has, to be sure, linkages of various kinds, arrangements with outside universities for certain graduate programs, and it adheres delicately, somewhat ambiguously, to the Woods Hole Oceanographic Institute just up the street. But it has never come under the domination of any outside institution or governmental agency, nor has it ever been told what to do by any outside group. Internally, the important institutional decisions seem to have been made by a process of accommodation and adaptation, with resistible forces always meeting movable objects.

The invertebrate eye was invented into an optical instrument at the MBL, opening the way to modern visual physiology. The giant axon of the Woods Hole squid became the apparatus for the creation of today's astonishing neurobiology. Developmental and reproductive biology were recognized and defined as sciences here, beginning with sea-urchin eggs and working up. Marine models were essential in the early days of research on muscle structure and function, and research on muscle has become a major preoccupation at the MBL. Ecology was a sober, industrious science here long ago, decades before the rest of us discovered the term. In recent years there have been expansion and strengthening in new fields; biologic membranes, immunology, genetics, and cell regulatory mechanisms are currently booming.

You can never tell when new things may be starting up from improbable lines of work. The amebocytes of starfish were recently found to contain a material that immobilizes the macrophages of mammals, resembling a product of immune lymphocytes in higher forms. Aplysia, a sea slug that looks as though it couldn't be good for anything, has been found by neurophysiologists to be filled with truth. Limulus, one of the world's conservative beasts, has recently been in the newspapers; it was discovered to contain a reagent for the detection of vanishingly small quantities of endotoxin from gram-negative bacteria, and the pharmaceutical industry has already sniffed commercial possibilities for the monitoring of pyrogen-free materials; horseshoe crabs may soon be as marketable as lobsters.

There is no way of predicting what the future will be like for an institution such as the MBL. One way or another, it will evolve. It may shift soon into a new phase, with a year-round program for teaching and research and a year-round staff, but it will have to accomplish this without jeopardizing the immense power of its summer programs, or all institutional hell will break loose. It will have to find new ways for relating to the universities, if its graduate programs are to expand as they should. It will have to develop new symbiotic relations with the Oceanographic Institute, since both places have so much at stake. And it will have to find more money, much more—the kind of money that only federal governments possess—without losing any of its own initiative.

It will be an interesting place to watch, in the years ahead. In a rational world, things ought to go as well for the MBL as they have in the past, and it should become an even larger and more agile collective intelligence. If you can think of good questions to ask about the life of the earth, it should be as good a place as any to go for answers.

It is now, in fact. You might begin at the local beach, which functions as a sort of ganglion. It is called Stony Beach, because it used to be covered, painfully, by small stones. Long ago, somehow, some committee of scientists, prodded by footsore wives, found enough money to cover it with a layer of sand. It is the most minor of beaches, hardly big enough for a committee, but close enough to the laboratories so that the investigators can walk down for a sandwich lunch with their children on sunny weekdays.

From time to time, pure physicists turn up, with only a few minutes to spare from a meeting at the National Academy summer headquarters, tired from making forecasts on classifiedly obscure matters, wearing the look of doom. The physicists are another species, whiter-skinned, towel-draped against the sun, unearthly, the soles of their feet so sensitive that they limp on sand.

A small boy, five-ish, with myopia and glasses, emerges from the water; characteristically, although his hair is dripping his glasses are bone dry; he has already begun to master technique. As he picks his way between the conversations, heading for his mother, who is explaining homology between DNA in chloroplasts and bacteria, he is shaking his head slowly in wonderment, looking at something brown and gelatinous held in his hand, saying, "That is very interesting water." At Stony Beach the water is regarded as primarily interesting, even by small boys.

On weekends, in hot midsummer, you can see how the governing mechanisms work. It is so crowded that one must pick one's way on tiptoe to find a hunching place, but there is always a lot of standing up anyway; biologists seem to prefer standing on beaches, talking at each other, gesturing to indicate the way things are assembled, bending down to draw diagrams in the sand. By the end of the day, the sand is crisscrossed with a mesh of ordinates, abscissas, curves to account for everything in nature.

You can hear the sound from the beach at a distance, before you see the people. It is that most extraordinary noise, half-shout, half-song, made by confluent, simultaneously raised human voices, explaining things to each other.

You hear a similar sound at the close of the Friday Evening Lecture, the MBL's weekly grand occasion, when the guest lecturers from around the world turn up to present their most stunning pieces of science. As the audience flows out of the auditorium, there is the same jubilant descant, the

great sound of crowded people explaining things to each
other as fast as their minds will work. You cannot make out
individual words in the mass, except that the recurrent
phrase, "But look—" keeps bobbing above the surf of language.

Not many institutions can produce this spontaneous music at will, summer after summer, year after year. It takes a
special gift, and the MBL appears to have been born with
it. Perhaps this is an aspect of the way we build language
after all. The scale is very small, and it is not at all clear how
it works, but it makes a nice thought for a time when we
can't seem to get anything straight or do anything right.

Autonomy

Working a typewriter by touch, like riding a bicycle or strolling on a path, is best done by not giving it a glancing thought. Once you do, your fingers fumble and hit the wrong keys. To do things involving practiced skills, you need to turn loose the systems of muscles and nerves responsible for each maneuver, place them on their own, and stay out of it. There is no real loss of authority in this, since you get to decide whether to do the thing or not, and you can intervene and embellish the technique any time you like; if you want to ride a bicycle backward, or walk with an eccentric loping gait giving a little skip every fourth step, whistling at the same time, you can do that. But if you concentrate your attention on the details, keeping in touch with each muscle, thrusting yourself into a free fall with each step and catching yourself at the last moment by sticking out the other foot in time to break the fall, you will end up immobilized, vibrating with fatigue.

It is a blessing to have options for choice and change in the learning of such unconsciously coordinated acts. If we were born with all these knacks inbuilt, automated like ants, we would surely miss the variety. It would be a less interesting world if we all walked and skipped alike, and never fell from bicycles. If we were all genetically programmed to play the piano deftly from birth, we might never learn to understand music.

The rules are different for the complicated, coordinated, fantastically skilled manipulations we perform with our insides. We do not have to learn anything. Our smooth-muscle cells are born with complete instructions, in need of no help from us, and they work away on their own schedules, modulating the lumen of blood vessels, moving things through intestines, opening and closing tubules according to the requirements of the entire system. Secretory cells elaborate their products in privacy; the heart contracts and relaxes; hormones are sent off to react silently with cell membranes, switching adenyl cyclase, prostaglandin, and other signals on and off; cells communicate with each other by simply touching; organelles send messages to other organelles; all this goes on continually, without ever a personal word from us. The arrangement is that of an ecosystem, with the operation of each part being governed by the state and function of all the other parts. When things are going well, as they generally are, it is an infallible mechanism.

But now the autonomy of this interior domain, long regarded as inviolate, is open to question. The experimental psychologists have recently found that visceral organs can be taught to do various things, as easily as a boy learns to ride a bicycle, by the instrumental techniques of operant conditioning. If a thing is done in the way the teacher wants, at a signal, and a suitable reward given immediately to reinforce the action, it becomes learned. Rats, rewarded by stimulation of their cerebral "pleasure centers," have been instructed to speed up or slow down their hearts at a signal, or to alter their blood pressures, or switch off certain waves in their electroencephalograms and switch on others.

The same technology has been applied to human beings, with other kinds of rewards, and the results have been startling. It is claimed that you can teach your kidneys to change the rate of urine formation, raise or lower your blood pressure, change your heart rate, write different brain waves, at will.

There is already talk of a breakthrough in the prevention

and treatment of human disease. According to proponents, when the technology is perfected and extended it will surely lead to new possibilities for therapy. If a rat can be trained to dilate the blood vessels of one of his ears more than those of the other, as has been reported, what rich experiences in self-control and self-operation may lie just ahead for man? There are already cryptic advertisements in the Personal columns of literary magazines, urging the purchase of electronic headsets for the training and regulation of one's own brain waves, according to one's taste.

You can have it.

Not to downgrade it. It is extremely important, I know, and one ought to feel elated by the prospect of taking personal charge, calling the shots, running one's cells around like toy trains. Now that we know that viscera can be taught, the thought comes naturally that we've been neglecting them all these years, and by judicious application of human intelligence, these primitive structures can be trained to whatever standards of behavior we wish to set for them.

My trouble, to be quite candid, is a lack of confidence in myself. If I were informed tomorrow that I was in direct communication with my liver, and could now take over, I would become deeply depressed. I'd sooner be told, forty thousand feet over Denver, that the 747 jet in which I had a coach seat was now mine to operate as I pleased; at least I would have the hope of bailing out, if I could find a parachute and discover quickly how to open a door. Nothing would save me and my liver, if I were in charge. For I am, to face the facts squarely, considerably less intelligent than my liver. I am, moreover, constitutionally unable to make hepatic decisions, and I prefer not to be obliged to, ever. I would not be able to think of the first thing to do.

I have the same feeling about the rest of my working parts. They are all better off without my intervention, in whatever they do. It might be something of a temptation to take over my brain, on paper, but I cannot imagine doing so in real life. I would lose track, get things mixed up, turn on wrong cells at wrong times, drop things. I doubt if I would ever be able to think up my own thoughts. My cells were born, or differentiated anyway, knowing how to do this kind of thing together. If I moved in to organize them they would resent it, perhaps become frightened, perhaps swarm out into my ventricles like bees.

Although it is, as I say, a temptation. I have never really been satisfied with the operation of my brain, and it might be fun to try running it myself, just once. There are several things I would change, given the opportunity: certain memories that tend to slip away unrecorded, others I've had enough of and would prefer to delete, certain notions I'd just as soon didn't keep popping in, trains of thought that go round and round without getting anywhere, rather like this one. I've always suspected that some of the cells in there are fluffing off much of the time, and I'd like to see a little more attention and real work. Also, while I'm about it, I could do with a bit more respect.

On balance, however, I think it best to stay out of this business. Once you began, there would be no end to the responsibilities. I'd rather leave all my automatic functions with as much autonomy as they please, and hope for the best. Imagine having to worry about running leukocytes, keeping track, herding them here and there, listening for signals. After the first flush of pride in ownership, it would be exhausting and debilitating, and there would be no time for anything else.

What to do, then? It cannot simply be left there. If we have learned anything at all in this century, it is that all new technologies will be put to use, sooner or later, for better or worse, as it is in our nature to do. We cannot expect an exception for the instrumental conditioning of autonomic functions. We will be driven to make use of it, trying to communicate with our internal environment, to meddle, and it will consume so much of our energy that we will end up even more cut off from things outside, missing the main sources of the sensation of living.

I have a suggestion for a way out. Given the capacity to control autonomic functions, modulate brain waves, run cells, why shouldn't it be possible to employ exactly the same technology to go in precisely the opposite direction? Instead of getting in there and taking things over, couldn't we learn to disconnect altogether, uncouple, detach, and float free? You would only need to be careful, if you tried it, that you let go of the right end.

Of course, people have been trying to do this sort of thing for a long time, by other techniques and with varying degrees of luck. This is what Zen archery seems to be about, come to think of it. You learn, after long months of study under a master, to release the arrow without releasing it yourself. Your fingers must do the releasing, on their own, remotely, like the opening of a flower. When you have learned this, no matter where the arrow goes, you have it made. You can step outside for a look around.

Organelles
as Organisms

We seem to be living through the biologic revolution, so far anyway, without being upheaved or even much disturbed by it. Even without being entirely clear about just what it is, we are all learning to take it for granted. It is a curious, peaceful sort of revolution, in which there is no general apprehension that old views are being outraged and overturned. Instead, whole, great new blocks of information are being brought in almost daily and put precisely down in what were previously empty spaces. The news about DNA and the genetic code did not displace an earlier dogma; there was nothing much there to be moved aside. Molecular biology did not drive out older, fixed views about the intimate details of cell function. We seem to be starting at the beginning, from scratch.

We not only take it for granted—we tend to talk about the biologic revolution as though expecting to make profits from it, rather like a version of last century's industrial revolution. All sorts of revolutionary changes in technology are postulated for the future, ranging from final control of human disease to solutions of the world food and population problems. We are even beginning to argue about which futures we like and which we prefer to cancel. Questions about the merits of genetic engineering, the cloning of desirable human beings from single cells, and even, I suppose, the possibility that two heads might actually be better than one, are already being debated at seminars.

So far, we don't seem to have been really shocked by anything among the items of new knowledge. There is surprise, even astonishment, but not yet dismay. Perhaps it is still too early to expect this, and it may lie just ahead.

It is not too early to begin looking for trouble. I can sense some, for myself anyway, in what is being learned about organelles. I was raised in the belief that these were obscure little engines inside my cells, owned and operated by me or my cellular delegates, private, submicroscopic bits of my intelligent flesh. Now, it appears, some of them, and the most important ones at that, are total strangers.

The evidence is strong, and direct. The membranes lining the inner compartment of mitochondria are unlike other animal cell membranes, and resemble most closely the membranes of bacteria. The DNA of mitochondria is qualitatively different from the DNA of animal cell nuclei and strikingly similar to bacterial DNA; moreover, like microbial DNA, it is closely associated with membranes. The RNA of mitochondria matches the organelles' DNA, but not that of the nucleus. The ribosomes inside the mitochondria are similar to bacterial ribosomes, and different from animal ribosomes. The mitochondria do not arise *de novo* in cells; they are always there, replicating on their own, independently of the replication of the cell. They travel down from egg to newborn; a few come in with the sperm, but most are maternal passengers.

The chloroplasts in all plants are, similarly, independent and self-replicating lodgers, with their own DNA and RNA and ribosomes. In structure and pigment content they are the images of prokaryotic blue-green algae. It has recently been reported that the nucleic acid of chloroplasts is, in fact, homologous with that of certain photosynthetic microorganisms.

There may be more. It has been suggested that flagellae and cilia were once spirochetes that joined up with the other

prokaryotes when nucleated cells were being pieced together. The centrioles and basal bodies are believed in some
quarters to be semiautonomous organisms with their own
separate genomes. Perhaps there are others, still unrecog-
nized.

I only hope I can retain title to my nuclei.

It is surprising that we take information like this so
calmly, as though it fitted in nicely with notions we've had
all along. Actually, the suggestion that chloroplasts and
mitochondria might be endosymbionts was made as long
ago as 1885, but one might expect, nevertheless, that con-
firmation of the suggestion would have sent the investigators out into the streets, hallooing. But this is a sober, industrious field, and the work goes on methodically, with special
interest just now in the molecular genetics of organelles.
There is careful, restrained speculation on how they got
there in the first place, with a consensus that they were
probably engulfed by larger cells more than a billion years
ago and have simply stayed there ever since.

The usual way of looking at them is as enslaved creatures,
captured to supply ATP for cells unable to respire on their
own, or to provide carbohydrate and oxygen for cells une-
quipped for photosynthesis. This master-slave arrangement
is the common view of full-grown biologists, eukaryotes all.
But there is the other side. From their own standpoint, the
organelles might be viewed as having learned early how to
have the best of possible worlds, with least effort and risk
to themselves and their progeny. Instead of evolving as we
have done, manufacturing longer and elaborately longer
strands of DNA, and running ever-increasing risks of mutat-
ing into evolutionary cul-de-sacs, they elected to stay small
and stick to one line of work. To accomplish this, and to
assure themselves the longest possible run, they got themselves inside all the rest of us.

It is a good thing for the entire enterprise that mito-

chondria and chloroplasts have remained small, conservative, and stable, since these two organelles are, in a fundamental sense, the most important living things on earth.
Between them they produce the oxygen and arrange for its
use. In effect, they run the place.

My mitochondria comprise a very large proportion of me.
I cannot do the calculation, but I suppose there is almost as
much of them in sheer dry bulk as there is the rest of me.
Looked at in this way, I could be taken for a very large,
motile colony of respiring bacteria, operating a complex
system of nuclei, microtubules, and neurons for the pleasure
and sustenance of their families, and running, at the moment, a typewriter.

I am intimately involved, and obliged to do a great deal
of essential work for my mitochondria. My nuclei code out
the outer membranes of each, and a good many of the
enzymes attached to the cristae must be synthesized by me.
Each of them, by all accounts, makes only enough of its own
materials to get along on, and the rest must come from me.
And I am the one who has to do the worrying.

Now that I know about the situation, I can find all kinds
of things to worry about. Viruses, for example. If my organelles are really symbiotic bacteria, colonizing me, what's
to prevent them from catching a virus, or if they have such
a thing as lysogeny, from conveying a phage to other organelles? Then there is the question of my estate. Do my
mitochondria all die with me, or did my children get some
of mine along with their mother's; this sort of thing should
not worry me, I know, but it does.

Finally, there is the whole question of my identity, and,
more than that, my human dignity. I did not mind it when
I first learned of my descent from lower forms of life. I had
in mind an arboreal family of beetle-browed, speechless,
hairy sub-men, ape-like, and I've never objected to them as
forebears. Indeed, being Welsh, I feel the better for it,

having clearly risen above them in my time of evolution. It is a source of satisfaction to be part of the improvement of the species.

But not these things. I had never bargained on descent from single cells without nuclei. I could even make my peace with that, if it were all, but there is the additional humiliation that I have not, in a real sense, descended at all. I have brought them all along with me, or perhaps they have brought me.

It is no good standing on dignity in a situation like this, and better not to try. It is a mystery. There they are, moving about in my cytoplasm, breathing for my own flesh, but strangers. They are much less closely related to me than to each other and to the free-living bacteria out under the hill. They feel like strangers, but the thought comes that the same creatures, precisely the same, are out there in the cells of sea gulls, and whales, and dune grass, and seaweed, and hermit crabs, and further inland in the leaves of the beech in my backyard, and in the family of skunks beneath the back fence, and even in that fly on the window. Through them, I am connected; I have close relatives, once removed, all over the place. This is a new kind of information, for me, and I regret somewhat that I cannot be in closer touch with my mitochondria. If I concentrate, I can imagine that I feel them; they do not quite squirm, but there is, from time to time, a kind of tingle. I cannot help thinking that if only I knew more about them, and how they maintain our synchrony, I would have a new way to explain music to myself.

There is something intrinsically good-natured about all symbiotic relations, necessarily, but this one, which is probably the most ancient and most firmly established of all, seems especially equable. There is nothing resembling predation, and no pretense of an adversary stance on either side. If you were looking for something like natural law to take the place of the "social Darwinism" of a century ago, you would have a hard time drawing lessons from the sense of life alluded to by chloroplasts and mitochondria, but there it is.

Germs

Watching television, you'd think we lived at bay, in total jeopardy, surrounded on all sides by human-seeking germs, shielded against infection and death only by a chemical technology that enables us to keep killing them off. We are instructed to spray disinfectants everywhere, into the air of our bedrooms and kitchens and with special energy into bathrooms, since it is our very own germs that seem the worst kind. We explode clouds of aerosol, mixed for good luck with deodorants, into our noses, mouths, underarms, privileged crannies—even into the intimate insides of our telephones. We apply potent antibiotics to minor scratches and seal them with plastic. Plastic is the new protector; we wrap the already plastic tumblers of hotels in more plastic, and seal the toilet seats like state secrets after irradiating them with ultraviolet light. We live in a world where the microbes are always trying to get at us, to tear us cell from cell, and we only stay alive and whole through diligence and fear.

We still think of human disease as the work of an organized, modernized kind of demonology, in which the bacteria are the most visible and centrally placed of our adversaries. We assume that they must somehow relish what they do. They come after us for profit, and there are so many of them that disease seems inevitable, a natural part of the human condition; if we succeed in eliminating one kind of disease there will always be a new one at hand, waiting to take its place.

These are paranoid delusions on a societal scale, explainable in part by our need for enemies, and in part by our memory of what things used to be like. Until a few decades ago, bacteria were a genuine household threat, and although most of us survived them, we were always aware of the nearness of death. We moved, with our families, in and out of death. We had lobar pneumonia, meningococcal meningitis, streptococcal infections, diphtheria, endocarditis, enteric fevers, various septicemias, syphilis, and, always, everywhere, tuberculosis. Most of these have now left most of us, thanks to antibiotics, plumbing, civilization, and money, but we remember.

In real life, however, even in our worst circumstances we have always been a relatively minor interest of the vast microbial world. Pathogenicity is not the rule. Indeed, it occurs so infrequently and involves such a relatively small number of species, considering the huge population of bacteria on the earth, that it has a freakish aspect. Disease usually results from inconclusive negotiations for symbiosis, an overstepping of the line by one side or the other, a biologic misinterpretation of borders.

Some bacteria are only harmful to us when they make exotoxins, and they only do this when they are, in a sense, diseased themselves. The toxins of diphtheria bacilli and streptococci are produced when the organisms have been infected by bacteriophage; it is the virus that provides the code for toxin. Uninfected bacteria are uninformed. When we catch diphtheria it is a virus infection, but not of us. Our involvement is not that of an adversary in a straightforward game, but more like blundering into someone else's accident.

I can think of a few microorganisms, possibly the tubercle bacillus, the syphilis spirochete, the malarial parasite, and a few others, that have a selective advantage in their ability to infect human beings, but there is nothing to be gained, in an evolutionary sense, by the capacity to cause illness or death. Pathogenicity may be something of a disadvantage for most microbes, carrying lethal risks more frightening to them than to us. The man who catches a meningococcus is in considerably less danger for his life, even without chemotherapy, than meningococci with the bad luck to catch a man. Most meningococci have the sense to stay out on the surface, in the rhinopharynx. During epidemics this is where they are to be found in the majority of the host population, and it generally goes well. It is only in the unaccountable minority, the "cases," that the line is crossed, and then there is the devil to pay on both sides, but most of all for the meningococci.

Staphylococci live all over us, and seem to have adapted to conditions in our skin that are uncongenial to most other bacteria. When you count them up, and us, it is remarkable how little trouble we have with the relation. Only a few of us are plagued by boils, and we can blame a large part of the destruction of tissues on the zeal of our own leukocytes. Hemolytic streptococci are among our closest intimates, even to the extent of sharing antigens with the membranes of our muscle cells; it is our reaction to their presence, in the form of rheumatic fever, that gets us into trouble. We can carry brucella for long periods in the cells of our reticuloendothelial system without any awareness of their existence; then cyclically, for reasons not understood but probably related to immunologic reactions on our part, we sense them, and the reaction of sensing is the clinical disease.

Most bacteria are totally preoccupied with browsing, altering the configurations of organic molecules so that they become usable for the energy needs of other forms of life. They are, by and large, indispensable to each other, living in interdependent communities in the soil or sea. Some have become symbionts in more specialized, local relations, living as working parts in the tissues of higher organisms. The root nodules of legumes would have neither form nor function without the masses of rhizobial bacteria swarming into root hairs, incorporating themselves with such intimacy that only an electron microscope can detect which membranes are bacterial and which plant. Insects have colonies of bacteria, the mycetocytes, living in them like little glands, doing heaven knows what but being essential. The microfloras of animal intestinal tracts are part of the nutritional system. And then, of course, there are the mitochondria and chloroplasts, permanent residents in everything.

The microorganisms that seem to have it in for us in the worst way—the ones that really appear to wish us ill—turn out on close examination to be rather more like bystanders, strays, strangers in from the cold. They will invade and replicate if given the chance, and some of them will get into our deepest tissues and set forth in the blood, but it is our response to their presence that makes the disease. Our arsenals for fighting off bacteria are so powerful, and involve so many different defense mechanisms, that we are in more danger from them than from the invaders. We live in the midst of explosive devices; we are mined.

It is the information carried by the bacteria that we cannot abide.

The gram-negative bacteria are the best examples of this. They display lipopolysaccharide endotoxin in their walls, and these macromolecules are read by our tissues as the very worst of bad news. When we sense lipopolysaccharide, we are likely to turn on every defense at our disposal; we will bomb, defoliate, blockade, seal off, and destroy all the tissues in the area. Leukocytes become more actively phagocytic, release lysosomal enzymes, turn sticky, and aggregate together in dense masses, occluding capillaries and shutting off the blood supply. Complement is switched on at the right point in its sequence to release chemotactic signals, calling in leukocytes from everywhere. Vessels become hyperreactive to epinephrine so that physiologic concentrations suddenly possess necrotizing properties. Pyrogen is released from leukocytes, adding fever to,hemorrhage, necrosis, and shock. It is a shambles.

All of this seems unnecessary, panic-driven. There is nothing intrinsically poisonous about endotoxin, but it must look awful, or feel awful, when sensed by cells. Cells believe that it signifies the presence of gram-negative bacteria, and they will stop at nothing to avoid this threat.

I used to think that only the most highly developed, civilized animals could be fooled in this way, but it is not so. The horseshoe crab is a primitive fossil of a beast, ancient and uncitified, but he is just as vulnerable to disorganization by endotoxin as a rabbit or a man. Bang has shown that an injection of a very small dose into the body cavity will cause the aggregation of hemocytes in ponderous, immovable masses that block the vascular channels, and a gelatinous clot brings the circulation to a standstill. It is now known that a limulus clotting system, perhaps ancestral to ours, is centrally involved in the reaction. Extracts of the hemocytes can be made to jell by adding extremely small amounts of endotoxin. The self-disintegration of the whole animal that follows a systemic injection can be interpreted as a well-intentioned but lethal error. The mechanism is itself quite a good one, when used with precision and restraint, admirably designed for coping with intrusion by a single bacterium: the hemocyte would be attracted to the site, extrude the coagulable protein, the microorganism would be entrapped and immobilized, and the thing would be finished. It is when confronted by the overwhelming signal of free molecules of endotoxin, evoking memories of vibrios in great numbers, that the limulus flies into panic, launches all his defenses at once, and destroys himself.

It is, basically, a response to propaganda, something like the panic-producing pheromones that slave-taking ants release to disorganize the colonies of their prey.

I think it likely that many of our diseases work in this way. Sometimes, the mechanisms used for overkill are immunologic, but often, as in the limulus model, they are more primitive kinds of memory. We tear ourselves to pieces because of symbols, and we are more vulnerable to this than to any host of predators. We are, in effect, at the mercy of our own Pentagons, most of the time.

Your Very Good Health

We spend $80 billion a year on health, as we keep reminding ourselves, or is it now $90 billion? Whichever, it is a shocking sum, and just to mention it is to suggest the presence of a vast, powerful enterprise, intricately organized and coordinated. It is, however, a bewildering, essentially scatterbrained kind of business, expanding steadily without being planned or run by anyone in particular. Whatever sum we spent last year was only discovered after we'd spent it, and nobody can be sure what next year's bill will be. The social scientists, attracted by problems of this magnitude, are beginning to swarm in from all quarters to take a closer look, and the economists are all over the place, pursing their lips and shaking their heads, shipping more and more data off to the computers, trying to decide whether this is a proper industry or a house of IBM cards. There doesn't seem to be any doubt about the amount of money being spent, but it is less certain where it goes, and for what.

It has become something of a convenience to refer to the whole endeavor as the "Health Industry." This provides the illusion that it is in a general way all one thing, and that it turns out, on demand, a single, unambiguous product, which is health. Thus, health care has become the new name for medicine. Health-care delivery is what doctors now do, along with hospitals and the other professionals who work with doctors, now known collectively as the health providers. The patients have become health consumers. Once you start on this line, there's no stopping. Just recently, to correct some of the various flaws, inequities, logistic defects, and near-bankruptcies in today's health-care delivery system, the government has officially invented new institutions called Health Maintenance Organizations, already known familiarly as HMO's, spreading out across the country like post offices, ready to distribute in neat packages, as though from a huge, newly stocked inventory, health.

Sooner or later, we are bound to get into trouble with this word. It is too solid and unequivocal a term to be used as a euphemism and this seems to be what we are attempting. I am worried that we may be overdoing it, taxing its meaning, to conceal an unmentionable reality that we've somehow agreed not to talk about in public. It won't work. Illness and death still exist and cannot be hidden. We are still beset by plain diseases, and we do not control them; they are loose on their own, afflicting us unpredictably and haphazardly. We are only able to deal with them when they have made their appearance, and we must use the methods of medical care for this, as best we can, for better or worse.

It would be a better world if this were not true, but the fact is that diseases do not develop just because of carelessness about the preservation of health. We do not become sick only because of a failure of vigilance. Most illnesses, especially the major ones, are blind accidents that we have no idea how to prevent. We are really not all that good at preventing disease or preserving health—not yet anyway—and we are not likely to be until we have learned a great deal about disease mechanisms.

There is disagreement on this point, of course. Some of the believers among us are convinced that once we get a health-care delivery system that really works, the country might become a sort of gigantic spa, offering, like the labels on European mineral-water bottles, preventives for everything from weak kidneys to moroseness.

It is a surprise that we haven't already learned that the word is a fallible incantation. Several decades of mental health have not made schizophrenia go away, nor has it been established that a community mental-health center can yet maintain the mental health of a community. These admirable institutions are demonstrably useful for the management of certain forms of mental disease, but that is another matter.

My complaint about the terms is that they sound too much like firm promises. A Health Maintenance Organization, if well organized and financed, will have the best features of a clinic and hospital and should be of value to any community, but the people will expect it to live up to its new name. It will become, with the sign over its door, an official institution for the distribution of health, and if intractable heart disease develops in anyone thereafter, as it surely will (or multiple sclerosis, or rheumatoid arthritis, or the majority of cancers that can neither be prevented nor cured, or chronic nephritis, or stroke, or moroseness), the people will begin looking sidelong and asking questions in a low voice.

Meanwhile, we are paying too little attention, and respect, to the built-in durability and sheer power of the human organism. Its surest tendency is toward stability and balance. It is a distortion, with something profoundly disloyal about it, to picture the human being as a teetering, fallible contraption, always needing watching and patching, always on the verge of flapping to pieces; this is the doctrine that people hear most often, and most eloquently, on all our information media. We ought to be developing a much better system for general education about human health, with more curricular time for acknowledgment, and even some celebration, of the absolute marvel of good health that is the real lot of most of us, most of the time.

The familiar questions about the needs of the future in

medicine are still before us. What items should be available, optimally, in an ideal health-care delivery system? How do you estimate the total need, per patient per year, for doctors, nurses, drugs, laboratory tests, hospital beds, x-rays, and so forth, in the best of rational worlds? My suggestion for a new way to develop answers is to examine, in detail, the ways in which the various parts of today's medical-care technology are used, from one day to the next, by the most sophisticated, knowledgeable, and presumably satisfied consumers who now have full access to the system—namely, the well-trained, experienced, middle-aged, married-with-family internists.

I could design the questionnaire myself, I think. How many times in the last five years have the members of your family, including yourself, had any kind of laboratory test? How many complete physical examinations? X-rays? Electrocardiograms? How often, in a year's turning, have you prescribed antibiotics of any kind for yourself or your family? How many hospitalizations? How much surgery? How many consultations with a psychiatrist? How many formal visits to a doctor, any doctor, including yourself?

I will bet that if you got this kind of information, and added everything up, you would find a quite different set of figures from the ones now being projected in official circles for the population at large. I have tried it already, in an unscientific way, by asking around among my friends. My data, still soft but fairly consistent, reveal that none of my internist friends have had a routine physical examination since military service; very few have been x-rayed except by dentists; almost all have resisted surgery; laboratory tests for anyone in the family are extremely rare. They use a lot of aspirin, but they seem to write very few prescriptions and almost never treat family fever with antibiotics. This is not to say that they do not become ill; these families have the same incidence of chiefly respiratory and gastrointestinal illness as everyone else, the same number of anxieties and bizarre notions, and the same number—on balance, a small number—of frightening or devastating diseases.

It will be protested that internists and their households are really full-time captive patients and cannot fairly be compared to the rest of the population. As each member of the family appears at the breakfast table, the encounter is, in effect, a house-call. The father is, in the liveliest sense, a family doctor. This is true, but all the more reason for expecting optimal use to be made of the full range of medicine's technology. There is no problem of access, the entire health-care delivery system is immediately at hand, and the cost of all items is surely less than that for nonmedical families. All the usual constraints that limit the use of medical care by the general population are absent.

If my hunch, based on the small sample of professional friends, is correct, these people appear to use modern medicine quite differently from the ways in which we have systematically been educating the public over the last few decades. It cannot be explained away as an instance of shoemakers' children going without shoes. Doctors' families do tend to complain that they receive less medical attention than their friends and neighbors, but they seem a normal, generally healthy lot, with a remarkably low incidence of iatrogenic illness.

The great secret, known to internists and learned early in marriage by internists' wives, but still hidden from the general public, is that most things get better by themselves. Most things, in fact, are better by morning.

It is conceivable that we might be able to provide good medical care for everyone needing it, in a new system designed to assure equity, provided we can restrain ourselves, or our computers, from designing a system in which all 200 million of us are assumed to be in constant peril of failed health every day of our lives. In the same sense that our judicial system presumes us to be innocent until proved guilty, a medical-care system may work best if it starts with the presumption that most people are healthy. Left to themselves, computers may try to do it in the opposite way, taking it as given that some sort of direct, continual, professional intervention is required all the time to maintain the health of each citizen, and we will end up spending all our money on nothing but that. Meanwhile, there is a long list of other things to do if we are to change the way we live together, especially in our cities, in time. Social health is another kind of problem, more complex and urgent, and there will be other bills to pay.

Social Talk

Not all social animals are social with the same degree of commitment. In some species, the members are so tied to each other and interdependent as to seem the loosely conjoined cells of a tissue. The social insects are like this; they move, and live all their lives, in a mass; a beehive is a spherical animal. In other species, less compulsively social, the members make their homes together, pool resources, travel in packs or schools, and share the food, but any single one can survive solitary, detached from the rest. Others are social only in the sense of being more or less congenial, meeting from time to time in committees, using social gatherings as *ad hoc* occasions for feeding and breeding. Some animals simply nod at each other in passing, never reaching even a first-name relationship.

It is not a simple thing to decide where we fit, for at one time or another in our lives we manage to organize in every imaginable social arrangement. We are as interdependent, especially in our cities, as bees or ants, yet we can detach if we wish and go live alone in the woods, in theory anyway. We feed and look after each other, constructing elaborate systems for this, even including vending machines to dispense ice cream in gas stations, but we also have numerous books to tell us how to live off the land. We cluster in family groups, but we tend, unpredictably, to turn on each other and fight as if we were different species. Collectively, we hanker to accumulate all the information in the universe and distribute it around among ourselves as though it were a kind of essential foodstuff, ant-fashion (the faintest trace of real news in science has the action of a pheromone, lifting

the hairs of workers in laboratories at the ends of the earth),
but each of us also builds a private store of his own secret
knowledge and hides it away like untouchable treasure. We
have names to label each as self, and we believe without
reservation that this system of taxonomy will guarantee the
entity, the absolute separateness of each of us, but the mechanism has no discernible function in the center of a crowded
city; we are essentially nameless, most of our time.

Nobody wants to think that the rapidly expanding mass
of mankind, spreading out over the surface of the earth,
blackening the ground, bears any meaningful resemblance
to the life of an anthill or a hive. Who would consider for
a moment that the more than 3 billion of us are a sort of
stupendous animal when we become linked together? We
are not mindless, nor is our day-to-day behavior coded out
to the last detail by our genomes, nor do we seem to be
engaged together, compulsively, in any single, universal,
stereotyped task analogous to the construction of a nest. If
we were ever to put all our brains together in fact, to make
a common mind the way the ants do, it would be an unthinkable thought, way over our heads.

Social animals tend to keep at a particular thing, generally
something huge for their size; they work at it ceaselessly
under genetic instructions and genetic compulsion, using it
to house the species and protect it, assuring permanence.

There are, to be sure, superficial resemblances in some of
the things we do together, like building glass and plastic
cities on all the land and farming under the sea, or assem-
bling in armies, or landing samples of ourselves on the
moon, or sending memoranda into the next galaxy. We do
these together without being quite sure why, but we can
stop doing one thing and move to another whenever we
like. We are not committed or bound by our genes to stick
to one activity forever, like the wasps. Today's behavior is
no more fixed than when we tumbled out over Europe to
build cathedrals in the twelfth century. At that time we were

convinced that it would go on forever, that this was the way to live, but it was not; indeed, most of us have already forgotten what it was all about. Anything we do in this transient, secondary social way, compulsively and with all our energies but only for a brief period of our history, cannot be counted as social behavior in the biological sense. If we can turn it on and off, on whims, it isn't likely that our genes are providing the detailed instructions. Constructing Chartres was good for our minds, but we found that our lives went on, and it is no more likely that we will find survival in Rome plows or laser bombs, or rapid mass transport or a Mars lander, or solar power, or even synthetic protein. We do tend to improvise things like this as we go along, but it is clear that we can pick and choose.

For practical purposes, it would probably be best for us not to be biologically social, in the long run. Not that we have a choice, or course, or even a vote. It would not be good news to learn that we are all roped together intellectually, droning away at some featureless, genetically driven collective work, building something so immense that we can never see the outlines. It seems especially hard, even perilous, for this to be the burden of a species with the unique attribute of speech, and argument. Leave this kind of life to the insects and birds, and lesser mammals, and fish.

But there is just that one thing. About human speech.

It begins to look, more and more disturbingly, as if the gift of language is the single human trait that marks us all genetically, setting us apart from all the rest of life. Language is, like nest-building or hive-making, the universal and biologically specific activity of human beings. We engage in it communally, compulsively, and automatically. We cannot be human without it; if we were to be separated from it our minds would die, as surely as bees lost from the hive.

We are born knowing how to use language. The capacity to recognize syntax, to organize and deploy words into intelligible sentences, is innate in the human mind. We are programmed to identify patterns and generate grammar. There are invariant and variable structures in speech that are common to all of us. As chicks are endowed with an innate capacity to read information in the shapes of over-hanging shadows, telling hawk from other birds, we can identify the meaning of grammar in a string of words, and we are born this way. According to Chomsky, who has examined it as a biologist looks at live tissue, language "must simply be a biological property of the human mind." The universal attributes of language are genetically set; we do not learn them, or make them up as we go along.

We work at this all our lives, and collectively we give it life, but we do not exert the least control over language, not as individuals or committees or academies or governments. Language, once it comes alive, behaves like an active, motile organism. Parts of it are always being changed, by a ceaseless activity to which all of us are committed; new words are invented and inserted, old ones have their meaning altered or abandoned. New ways of stringing words and sentences together come into fashion and vanish again, but the underlying structure simply grows, enriches itself, expands. Individual languages age away and seem to die, but they leave progeny all over the place. Separate languages can exist side by side for centuries without touching each other, maintaining their integrity with the vigor of incompatible tissues. At other times, two languages may come together, fuse, replicate, and give rise to nests of new tongues.

If language is at the core of our social existence, holding us together, housing us in meaning, it may also be safe to say that art and music are functions of the same universal, genetically determined mechanism. These are not bad things to do together. If we are social creatures because of this, and therefore like ants, I for one (or should I say we for one?) do not mind.

Information

According to the linguistic school currently on top, human beings are all born with a genetic endowment for recognizing and formulating language. This must mean that we possess genes for all kinds of information, with strands of special, peculiarly human DNA for the discernment of meaning in syntax. We must imagine the morphogenesis of deep structures, built into our minds, for coding out, like proteins, the parts of speech. Correct grammar (correct in the logical, not fashionable, sense) is as much a biologic characteristic of our species as feathers on birds.

If this is true, it would mean that the human mind is preset, in some primary sense, to generate more than just the parts of speech. Since everything else that we recognize as human behavior derives from the central mechanism of language, the same sets of genes are at least indirectly responsible for governing such astonishing behavior as in the concert hall, where hundreds of people crowd together, silent, head-tilted, meditating, listening to music as though receiving instructions, or in a gallery, moving along slowly, peering, never looking at each other, concentrating as though reading directions.

This view of things is compatible with the very old notion that a framework for meaning is somehow built into our minds at birth. We start our lives with templates, and attach to them, as we go along, various things that fit. There are neural centers for generating, spontaneously, numberless hypotheses about the facts of life. We store up information the way cells store energy. When we are lucky enough to find a direct match between a receptor and a fact, there is a deep explosion in the mind; the idea suddenly enlarges, rounds up, bursts with new energy, and begins to replicate. At times there are chains of reverberating explosions, shaking everything: the imagination, as we say, is staggered.

This system seems to be restricted to human beings, since we are the only beings with language, although chimpanzees may have the capability of manipulating symbols with a certain syntax. The great difference between us and the other animals may be the qualitative difference made by speech. We live by making transformations of energy into words, storing it up, and releasing it in controlled explosions.

Speechless animals cannot do this sort of thing, and they are limited to single-stage transactions. They wander, as we do, searching for facts to fit their sparser stock of hypotheses, but when the receptor meets its match, there is only a single thud. Without language, the energy that is encoiled, springlike, inside information can only be used once. The solitary wasp, Sphex, nearing her time of eggs, travels aloft with a single theory about caterpillars. She is, in fact, a winged receptor for caterpillars. Finding one to match the hypothesis, she swoops, pins it, paralyzes it, carries it off, and descends to deposit it precisely in front of the door of the round burrow (which, obsessed by a different version of the same theory, she had prepared beforehand). She drops the beast, enters the burrow, inspects the interior for last-minute irregularities, then comes out to pull it in for the egg-laying. It has the orderly, stepwise look of a well thought-out business. But if, while she is inside inspecting, you move the caterpillar a short distance, she has a less sensible second thought about the matter. She emerges, searches for a moment, finds it, drags it back to the original spot, drops it again, and runs inside to check the burrow again. If you move the caterpillar again, she will repeat the program, and you can keep her totally preoccupied for as long as you have the patience and the heart for it. It is a compulsive, essentially neurotic kind of behavior, as mindless as an Ionesco character, but the wasp cannot imagine any other way of doing the thing.

Lymphocytes, like wasps, are genetically programmed for exploration, but each of them seems to be permitted a different, solitary idea. They roam through the tissues, sensing and monitoring. Since there are so many of them, they can make collective guesses at almost anything antigenic on the surface of the earth, but they must do their work one notion at a time. They carry specific information in their surface receptors, presented in the form of a question: is there, anywhere out there, my particular molecular configuration? It seems to be in the nature of biologic information that it not only stores itself up as energy but also instigates a search for more. It is an insatiable mechanism.

Lymphocytes are apparently informed about everything foreign around them, and some of them come equipped for fitting with polymers that do not exist until organic chemists synthesize them in their laboratories. The cells can do more than predict reality; they are evidently programmed with wild guesses as well.

Not all animals have lymphocytes with the same range of information, as you might expect. As with language, the system is governed by genes, and there are genetic differences between species and between inbred animals of the same species. There are polymers that will fit the receptors of one line of guinea pigs or mice but not others; there are responders and nonresponders.

When the connection is made, and a particular lymphocyte with a particular receptor is brought into the presence of the particular antigen, one of the greatest small spectacles in nature occurs. The cell enlarges, begins making new DNA at a great rate, and turns into what is termed, appropriately, a blast. It then begins dividing, replicating itself into a new colony of identical cells, all labeled with the same receptor, primed with the same question. The new cluster is a memory, nothing less.

For this kind of mechanism to be useful, the cells are required to stick precisely to the point. Any ambiguity, any tendency to wander from the matter at hand, will introduce grave hazards for the cells, and even more for the host in which they live. Minor inaccuracies may cause reactions in which neighboring cells are recognized as foreign, and done in. There is a theory that the process of aging may be due to the cumulative effect of imprecision, a gradual degrading of information. It is not a system that allows for deviating.

Perhaps it is in this respect that language differs most sharply from other biologic systems for communication. Ambiguity seems to be an essential, indispensable element for the transfer of information from one place to another by words, where matters of real importance are concerned. It is often necessary, for meaning to come through, that there be an almost vague sense of strangeness and askewness. Speechless animals and cells cannot do this. The specifically locked-on antigen at the surface of a lymphocyte does not send the cell off in search of something totally different; when a bee is tracking sugar by polarized light, observing the sun as though consulting his watch, he does not veer away to discover an unimaginable marvel of a flower. Only the human mind is designed to work in this way, programmed to drift away in the presence of locked-on information, straying from each point in a hunt for a better, different point.

If it were not for the capacity for ambiguity, for the sensing of strangeness, that words in all languages provide, we would have no way of recognizing the layers of counterpoint in meaning, and we might be spending all our time sitting on stone fences, staring into the sun. To be sure, we would always have had some everyday use to make of the alphabet, and we might have reached the same capacity for small talk, but it is unlikely that we would have been able to evolve from words to Bach. The great thing about human language is that it prevents us from sticking to the matter at hand.

Death in the Open

ost of the dead animals you see on highways near the cities are dogs, a few cats. Out in the countryside, the forms and coloring of the dead are strange; these are the wild creatures. Seen from a car window they appear as fragments, evoking memories of woodchucks, badgers, skunks, voles, snakes, sometimes the mysterious wreckage of a deer.

It is always a queer shock, part a sudden upwelling of grief, part unaccountable amazement. It is simply astounding to see an animal dead on a highway. The outrage is more than just the location; it is the impropriety of such visible death, anywhere. You do not expect to see dead animals in the open. It is the nature of animals to die alone, off somewhere, hidden. It is wrong to see them lying out on the highway; it is wrong to see them anywhere.

Everything in the world dies, but we only know about it as a kind of abstraction. If you stand in a meadow, at the edge of a hillside, and look around carefully, almost everything you can catch sight of is in the process of dying, and most things will be dead long before you are. If it were not for the constant renewal and replacement going on before your eyes, the whole place would turn to stone and sand under your feet.

There are some creatures that do not seem to die at all; they simply vanish totally into their own progeny. Single cells do this. The cell becomes two, then four, and so on, and after a while the last trace is gone. It cannot be seen as death; barring mutation, the descendants are simply the first cell, living all over again. The cycles of the slime mold have episodes that seem as conclusive as death, but the withered slug, with its stalk and fruiting body, is plainly the transient tissue of a developing animal; the free-swimming amebocytes use this organ collectively in order to produce more of themselves.

There are said to be a billion billion insects on the earth at any moment, most of them with very short life expectancies by our standards. Someone has estimated that there are 25 million assorted insects hanging in the air over every temperate square mile, in a column extending upward for thousands of feet, drifting through the layers of the atmosphere like plankton. They are dying steadily, some by being eaten, some just dropping in their tracks, tons of them around the earth, disintegrating as they die, invisibly.

Who ever sees dead birds, in anything like the huge numbers stipulated by the certainty of the death of all birds? A dead bird is an incongruity, more startling than an unexpected live bird, sure evidence to the human mind that something has gone wrong. Birds do their dying off somewhere, behind things, under things, never on the wing.

Animals seem to have an instinct for performing death alone, hidden. Even the largest, most conspicuous ones find ways to conceal themselves in time. If an elephant missteps and dies in an open place, the herd will not leave him there; the others will pick him up and carry the body from place to place, finally putting it down in some inexplicably suitable location. When elephants encounter the skeleton of an elephant out in the open, they methodically take up each of the bones and distribute them, in a ponderous ceremony, over neighboring acres.

It is a natural marvel. All of the life of the earth dies, all of the time, in the same volume as the new life that dazzles us each morning, each spring. All we see of this is the odd stump, the fly struggling on the porch floor of the summer house in October, the fragment on the highway. I have lived all my life with an embarrassment of squirrels in my backyard, they are all over the place, all year long, and I have never seen, anywhere, a dead squirrel.

I suppose it is just as well. If the earth were otherwise, and all the dying were done in the open, with the dead there to be looked at, we would never have it out of our minds. We can forget about it much of the time, or think of it as an accident to be avoided, somehow. But it does make the process of dying seem more exceptional than it really is, and harder to engage in at the times when we must ourselves engage.

In our way, we conform as best we can to the rest of nature. The obituary pages tell us of the news that we are dying away, while the birth announcements in finer print, off at the side of the page, inform us of our replacements, but we get no grasp from this of the enormity of scale. There are 3 billion of us on the earth, and all 3 billion must be dead, on a schedule, within this lifetime. The vast mortality, involving something over 50 million of us each year, takes place in relative secrecy. We can only really know of the deaths in our households, or among our friends. These, detached in our minds from all the rest, we take to be unnatural events, anomalies, outrages. We speak of our own dead in low voices; struck down, we say, as though visible death can only occur for cause, by disease or violence, avoidably. We send off for flowers, grieve, make ceremonies, scatter bones, unaware of the rest of the 3 billion on the same schedule. All of that immense mass of flesh and bone and consciousness will disappear by absorption into the earth, without recognition by the transient survivors.

Less than a half century from now, our replacements will have more than doubled the numbers. It is hard to see how we can continue to keep the secret, with such multitudes doing the dying. We will have to give up the notion that death is catastrophe, or detestable, or avoidable, or even strange. We will need to learn more about the cycling of life in the rest of the system, and about our connection to the process. Everything that comes alive seems to be in trade for something that dies, cell for cell. There might be some comfort in the recognition of synchrony, in the information that we all go down together, in the best of company.

Natural Science

The essential wildness of science as a manifestation of human behavior is not generally perceived. As we extract new things of value from it, we also keep discovering parts of the activity that seem in need of better control, more efficiency, less unpredictability. We'd like to pay less for it and get our money's worth on some more orderly, businesslike schedule. The Washington planners are trying to be helpful in this, and there are new programs for the centralized organization of science all over the place, especially in the biomedical field.

It needs thinking about. There is an almost ungovernable, biologic mechanism at work in scientific behavior at its best, and this should not be overlooked.

The difficulties are more conspicuous when the problems are very hard and complicated and the facts not yet in. Solutions cannot be arrived at for problems of this sort until the science has been lifted through a preliminary, turbulent zone of outright astonishment. Therefore, what must be planned for, in the laboratories engaged in the work, is the totally unforeseeable. If it is centrally organized, the system must be designed primarily for the elicitation of disbelief and the celebration of surprise.

Moreover, the whole scientific enterprise must be arranged so that the separate imaginations in different human minds can be pooled, and this is more a kind of game than a systematic business. It is in the abrupt, unaccountable aggregation of random notions, intuitions, known in science as good ideas, that the high points are made.

The most mysterious aspect of difficult science is the way it is done. Not the routine, not just the fitting together of things that no one had guessed at fitting, not the making of connections; these are merely the workaday details, the methods of operating. They are interesting, but not as fascinating as the central mystery, which is that we do it at all, and that we do it under such compulsion.

I don't know of any other human occupation, even including what I have seen of art, in which the people engaged in it are so caught up, so totally preoccupied, so driven beyond their strength and resources.

Scientists at work have the look of creatures following genetic instructions; they seem to be under the influence of a deeply placed human instinct. They are, despite their efforts at dignity, rather like young animals engaged in savage play. When they are near to an answer their hair stands on end, they sweat, they are awash in their own adrenalin. To grab the answer, and grab it first, is for them a more powerful drive than feeding or breeding or protecting themselves against the elements.

It sometimes looks like a lonely activity, but it is as much the opposite of lonely as human behavior can be. There is nothing so social, so communal, so interdependent. An active field of science is like an immense intellectual anthill; the individual almost vanishes into the mass of minds tumbling over each other, carrying information from place to place, passing it around at the speed of light.

There are special kinds of information that seem to be chemotactic. As soon as a trace is released, receptors at the back of the neck are caused to tremble, there is a massive convergence of motile minds flying upwind on a gradient of surprise, crowding around the source. It is an infiltration of intellects, an inflammation.

There is nothing to touch the spectacle. In the midst of what seems a collective derangement of minds in total disorder, with bits of information being scattered about, torn to shreds, disintegrated, reconstituted, engulfed, in a kind of activity that seems as random and agitated as that of bees in a disturbed part of the hive, there suddenly emerges, with the purity of a slow phrase of music, a single new piece of truth about nature.

In short, it works. It is the most powerful and productive of the things human beings have learned to do together in many centuries, more effective than farming, or hunting and fishing, or building cathedrals, or making money.

It is instinctive behavior, in my view, and I do not understand how it works. It cannot be prearranged in any precise way; the minds cannot be lined up in tidy rows and given directions from printed sheets. You cannot get it done by instructing each mind to make this or that piece, for central committees to fit with the pieces made by other instructed minds. It does not work this way.

What it needs is for the air to be made right. If you want a bee to make honey, you do not issue protocols on solar navigation or carbohydrate chemistry, you put him together with other bees (and you'd better do this quickly, for solitary bees do not stay alive) and you do what you can to arrange the general environment around the hive. If the air is right, the science will come in its own season, like pure honey.

There is something like aggression in the activity, but it differs from other forms of aggressive behavior in having no sort of destruction as the objective. While it is going on, it looks and feels like aggression: get at it, uncover it, bring it out, grab it, it's mine! It is like a primitive running hunt, but there is nothing at the end of it to be injured. More probably, the end is a sigh. But then, if the air is right and the science is going well, the sigh is immediately interrupted, there is a yawping new question, and the wild, tumbling activity begins once more, out of control all over again.

Natural Man

The social scientists, especially the economists, are moving deeply into ecology and the environment these days, with disquieting results. It goes somehow against the grain to learn that cost-benefit analyses can be done neatly on lakes, meadows, nesting gannets, even whole oceans. It is hard enough to confront the environmental options ahead, and the hard choices, but even harder when the price tags are so visible. Even the new jargon is disturbing: it hurts the spirit, somehow, to read the word *environments,* when the plural means that there are so many alternatives there to be sorted through, as in a market, and voted on. Economists need cool heads and cold hearts for this sort of work, and they must write in icy, often skiddy prose.

The degree to which we are all involved in the control of the earth's life is just beginning to dawn on most of us, and it means another revolution for human thought.

This will not come easily. We've just made our way through inconclusive revolutions on the same topic, trying to make up our minds how we feel about nature. As soon as we arrived at one kind of consensus, like an enormous committee, we found it was time to think it through all over, and now here we are, at it again.

The oldest, easiest-to-swallow idea was that the earth was man's personal property, a combination of garden, zoo, bank vault, and energy source, placed at our disposal to be consumed, ornamented, or pulled apart as we wished. The betterment of mankind was, as we understood it, the whole point of the thing. Mastery over nature, mystery and all, was a moral duty and social obligation.

In the last few years we were wrenched away from this way of looking at it, and arrived at something like general agreement that we had it wrong. We still argue the details, but it is conceded almost everywhere that we are not the masters of nature that we thought ourselves; we are as dependent on the rest of life as are the leaves or midges or fish. We are part of the system. One way to put it is that the earth is a loosely formed, spherical organism, with all its working parts linked in symbiosis. We are, in this view, neither owners nor operators; at best, we might see ourselves as motile tissue specialized for receiving information—perhaps, in the best of all possible worlds, functioning as a nervous system for the whole being.

There is, for some, too much dependency in this view, and they prefer to see us as a separate, qualitatively different, special species, unlike any other form of life, despite the sharing around of genes, enzymes, and organelles. No matter, there is still the underlying idea that we cannot have a life of our own without concern for the ecosystem in which we live, whether in majesty or not. This idea has been strong enough to launch the new movements for the sustenance of wilderness, the protection of wildlife, the turning off of insatiable technologies, the preservation of "whole earth."

But now, just when the new view seems to be taking hold, we may be in for another wrench, this time more dismaying and unsettling than anything we've come through. In a sense, we shall be obliged to swing back again, still believing in the new way but constrained by the facts of life to live in the old. It may be too late, as things have turned out.

We are, in fact, the masters, like it or not.

It is a despairing prospect. Here we are, practically speaking twenty-first-century mankind, filled to exuberance with our new understanding of kinship to all the family of life, and here we are, still nineteenth-century man, walking boot-shod over the open face of nature, subjugating and civilizing it. And we cannot stop this controlling, unless we vanish under the hill ourselves. If there were such a thing as a world mind, it should crack over this.

The truth is, we have become more deeply involved than we ever dreamed. The fact that we sit around as we do, worrying seriously about how best to preserve the life of the earth, is itself the sharpest measure of our involvement. It is not human arrogance that has taken us in this direction, but the most natural of natural events. We developed this way, we grew this way, we are this kind of species.

We have become, in a painful, unwished-for way, nature itself. We have grown into everywhere, spreading like a new growth over the entire surface, touching and affecting every other kind of life, *incorporating* ourselves. The earth risks being eutrophied by us. We are now the dominant feature of our own environment. Humans, large terrestrial metazoans, fired by energy from microbial symbionts lodged in their cells, instructed by tapes of nucleic acid stretching back to the earliest live membranes, informed by neurons essentially the same as all the other neurons on earth, sharing structures with mastodons and lichens, living off the sun, are now in charge, running the place, for better or worse.

Or is it really this way? It could be, you know, just the other way around. Perhaps we are the invaded ones, the subjugated, used.

Certain animals in the sea live by becoming part-animal, part-plant. They engulf algae, which then establish themselves as complex plant tissues, essential for the life of the whole company. I suppose the giant clam, if he had more of a mind, would have moments of dismay on seeing what he has done to the plant world, incorporating so much of it, enslaving green cells, living off the photosynthesis. But the plant cells would take a different view of it, having captured the clam on the most satisfactory of terms, including the small lenses in his tissues that focus sunlight for their benefit; perhaps algae have bad moments about what they may collectively be doing to the world of clams.

With luck, our own situation might be similar, on a larger scale. This might turn out to be a special phase in the morphogenesis of the earth when it is necessary to have something like us, for a time anyway, to fetch and carry energy, look after new symbiotic arrangements, store up information for some future season, do a certain amount of ornamenting, maybe even carry seeds around the solar system. That kind of thing. Handyman for the earth.

I would much prefer this useful role, if I had any say, to the essentially unearthly creature we seem otherwise on the way to becoming. It would mean making some quite fundamental changes in our attitudes toward each other, if we were really to think of ourselves as indispensable elements of nature. We would surely become the environment to worry about the most. We would discover, in ourselves, the sources of wonderment and delight that we have discerned in all other manifestations of nature. Who knows, we might even acknowledge the fragility and vulnerability that always accompany high specialization in biology, and movements might start up for the protection of ourselves as a valuable, endangered species. We couldn't lose.

The Iks

The small tribe of Iks, formerly nomadic hunters and gatherers in the mountain valleys of northern Uganda, have become celebrities, literary symbols for the ultimate fate of disheartened, heartless mankind at large. Two disastrously conclusive things happened to them: the government decided to have a national park, so they were compelled by law to give up hunting in the valleys and become farmers on poor hillside soil, and then they were visited for two years by an anthropologist who detested them and wrote a book about them.

The message of the book is that the Iks have transformed themselves into an irreversibly disagreeable collection of unattached, brutish creatures, totally selfish and loveless, in response to the dismantling of their traditional culture. Moreover, this is what the rest of us are like in our inner selves, and we will all turn into Iks when the structure of our society comes all unhinged.

The argument rests, of course, on certain assumptions about the core of human beings, and is necessarily speculative. You have to agree in advance that man is fundamentally a bad lot, out for himself alone, displaying such graces as affection and compassion only as learned habits. If you take this view, the story of the Iks can be used to confirm it. These people seem to be living together, clustered in small, dense villages, but they are really solitary, unrelated individuals with no evident use for each other. They talk, but only to make ill-tempered demands and cold refusals. They share nothing. They never sing. They turn the children out to forage as soon as they can walk, and desert the elders to starve whenever they can, and the foraging children snatch food from the mouths of the helpless elders. It is a mean society.

They breed without love or even casual regard. They defecate on each other's doorsteps. They watch their neighbors for signs of misfortune, and only then do they laugh. In the book they do a lot of laughing, having so much bad luck. Several times they even laughed at the anthropologist, who found this especially repellent (one senses, between the lines, that the scholar is not himself the world's luckiest man). Worse, they took him into the family, snatched his food, defecated on his doorstep, and hooted dislike at him. They gave him two bad years.

It is a depressing book. If, as he suggests, there is only Ikness at the center of each of us, our sole hope for hanging on to the name of humanity will be in endlessly mending the structure of our society, and it is changing so quickly and completely that we may never find the threads in time. Meanwhile, left to ourselves alone, solitary, we will become the same joyless, zestless, untouching lone animals.

But this may be too narrow a view. For one thing, the Iks are extraordinary. They are absolutely astonishing, in fact. The anthropologist has never seen people like them anywhere, nor have I. You'd think, if they were simply examples of the common essence of mankind, they'd seem more recognizable. Instead, they are bizarre, anomalous. I have known my share of peculiar, difficult, nervous, grabby people, but I've never encountered any genuinely, consistently detestable human beings in all my life. The Iks sound more like abnormalities, maladies.

I cannot accept it. I do not believe that the Iks are representative of isolated, revealed man, unobscured by social habits. I believe their behavior is something extra, something laid on. This unremitting, compulsive repellence is a kind of complicated ritual. They must have learned to act this way; they copied it, somehow.

I have a theory, then. The Iks have gone crazy.

The solitary Ik, isolated in the ruins of an exploded culture, has built a new defense for himself. If you live in an unworkable society you can make up one of your own, and this is what the Iks have done. Each Ik has become a group, a one-man tribe on its own, a constituency.

Now everything falls into place. This is why they do seem, after all, vaguely familiar to all of us. We've seen them before. This is precisely the way groups of one size or another, ranging from committees to nations, behave. It is, of course, this aspect of humanity that has lagged behind the rest of evolution, and this is why the Ik seems so primitive. In his absolute selfishness, his incapacity to give anything away, no matter what, he is a successful committee. When he stands at the door of his hut, shouting insults at his neighbors in a loud harangue, he is city addressing another city.

Cities have all the Ik characteristics. They defecate on doorsteps, in rivers and lakes, their own or anyone else's. They leave rubbish. They detest all neighboring cities, give nothing away. They even build institutions for deserting elders out of sight.

Nations are the most Iklike of all. No wonder the Iks seem familiar. For total greed, rapacity, heartlessness, and irresponsibility there is nothing to match a nation. Nations, by law, are solitary, self-centered, withdrawn into themselves. There is no such thing as affection between nations, and certainly no nation ever loved another. They bawl insults from their doorsteps, defecate into whole oceans, snatch all the food, survive by detestation, take joy in the bad luck of others, celebrate the death of others, live for the death of others.

That's it, and I shall stop worrying about the book. It does not signify that man is a sparse, inhuman thing at his center. He's all right. It only says what we've always known and never had enough time to worry about, that we haven't yet learned how to stay human when assembled in masses. The Ik, in his despair, is acting out this failure, and perhaps we should pay closer attention. Nations have themselves become too frightening to think about, but we might learn some things by watching these people.

Computers

You can make computers that are almost human. In some respects they are superhuman; they can beat most of us at chess, memorize whole telephone books at a glance, compose music of a certain kind and write obscure poetry, diagnose heart ailments, send personal invitations to vast parties, even go transiently crazy. No one has yet programmed a computer to be of two minds about a hard problem, or to burst out laughing, but that may come. Sooner or later, there will be real human hardware, great whirring, clicking cabinets intelligent enough to read magazines and vote, able to think rings around the rest of us.

Well, maybe, but not for a while anyway. Before we begin organizing sanctuaries and reservations for our software selves, lest we vanish like the whales, here is a thought to relax with.

Even when technology succeeds in manufacturing a machine as big as Texas to do everything we recognize as human, it will still be, at best, a single individual. This amounts to nothing, practically speaking. To match what we can do, there would have to be 3 billion of them with more coming down the assembly line, and I doubt that anyone will put up the money, much less make room. And even so, they would all have to be wired together, intricately and delicately, as we are, communicating with each other, talking incessantly, listening. If they weren't *at* each other this way, all their waking hours, they wouldn't be anything like human, after all. I think we're safe, for a long time ahead.

It is in our collective behavior that we are most mysterious. We won't be able to construct machines like ourselves until we've understood this, and we're not even close. All we know is the phenomenon: we spend our time sending messages to each other, talking and trying to listen at the same time, exchanging information. This seems to be our most urgent biological function; it is what we do with our lives. By the time we reach the end, each of us has taken in a staggering store, enough to exhaust any computer, much of it incomprehensible, and we generally manage to put out even more than we take in. Information is our source of energy; we are driven by it. It has become a tremendous enterprise, a kind of energy system on its own. All 3 billion of us are being connected by telephones, radios, television sets, airplanes, satellites, harangues on public-address systems, newspapers, magazines, leaflets dropped from great heights, words got in edgewise. We are becoming a grid, a circuitry around the earth. If we keep at it, we will become a computer to end all computers, capable of fusing all the thoughts of the world into a syncytium.

Already, there are no closed, two-way conversations. Any word you speak this afternoon will radiate out in all directions, around town before tomorrow, out and around the world before Tuesday, accelerating to the speed of light, modulating as it goes, shaping new and unexpected messages, emerging at the end as an enormously funny Hungarian joke, a fluctuation in the money market, a poem, or simply a long pause in someone's conversation in Brazil.

We do a lot of collective thinking, probably more than any other social species, although it goes on in something like secrecy. We don't acknowledge the gift publicly, and we are not as celebrated as the insects, but we do it. Effortlessly, without giving it a moment's thought, we are capable of changing our language, music, manners, morals, entertainment, even the way we dress, all around the earth in a year's turning. We seem to do this by general agreement, without voting or even polling. We simply think our way along, pass information around, exchange codes disguised as art, change our minds, transform ourselves.

Computers cannot deal with such levels of improbability, and it is just as well. Otherwise, we might be tempted to take over the control of ourselves in order to make long-range plans, and that would surely be the end of us. It would mean that some group or other, marvelously intelligent and superbly informed, undoubtedly guided by a computer, would begin deciding what human society ought to be like, say, over the next five hundred years or so, and the rest of us would be persuaded, one way or another, to go along. The process of social evolution would then grind to a stand-still, and we'd be stuck in today's rut for a millennium.

Much better we work our way out of it on our own, without governance. The future is too interesting and dangerous to be entrusted to any predictable, reliable agency. We need all the fallibility we can get. Most of all, we need to preserve the absolute unpredictability and total improbability of our connected minds. That way we can keep open all the options, as we have in the past.

It would be nice to have better ways of monitoring what we're up to so that we could recognize change while it is occurring, instead of waking up as we do now to the astonished realization that the whole century just past wasn't what we thought it was, at all. Maybe computers can be used to help in this, although I rather doubt it. You can make simulation models of cities, but what you learn is that they seem to be beyond the reach of intelligent analysis; if you try to use common sense to make predictions, things get more botched up than ever. This is interesting, since a city is the most concentrated aggregation of humans, all exerting whatever influence they can bring to bear. The city seems to have a life of its own. If we cannot understand how this works, we are not likely to get very far with human society at large.

Still, you'd think there would be some way in. Joined

together, the great mass of human minds around the earth seems to behave like a coherent, living system. The trouble is that the flow of information is mostly one-way. We are all obsessed by the need to feed information in, as fast as we can, but we lack sensing mechanisms for getting anything much back. I will confess that I have no more sense of what goes on in the mind of mankind than I have for the mind of an ant. Come to think of it, this might be a good place to start.

The Planning
of Science

I t is generally accepted that the biologic sciences are absolutely splendid. In just the past decade, they have uncovered a huge mass of brand-new information, and there is plenty more ahead; the biologic revolution is evidently still in its early stages. Everyone approves. By contrast, the public view of the progress of medicine during the same period is restrained, qualified, a mixture of hope and worry. For all the new knowledge, we still have formidable diseases, still unsolved, lacking satisfactory explanation, lacking satisfactory treatment. Why, it is asked, does the supply of new miracle drugs lag so far behind, while biology continues to move from strength to strength, elaborating new, powerful technologies for explaining, in fine detail, the very processes of life?

It doesn't seem to help to apply the inclusive term "biomedical" to our science, much as we would like to show that we are all one field of inquiry, share and share alike. There is still the conspicuous asymmetry between molecular biology and, say, the therapy of lung cancer. We may as well face up to it: there is a highly visible difference between the pace of basic science and the application of new knowledge to human problems. It needs explaining.

This is an especially lively problem at the moment, because of the immediate implications for national science policy. It is administratively fashionable in Washington to attribute the delay of applied science in medicine to a lack of systematic planning. Under a new kind of management, it is said, with more businesslike attention to the invention of practical applications, we should arrive at our targets more quickly and, it is claimed as a bonus, more economically. Targeting is the new word. We need more targeted research, more mission-oriented science. And maybe less basic research—maybe considerably less. This is said to be the new drift.

One trouble with this view is that it attributes to biology and medicine a much greater store of usable information, with coherence and connectedness, than actually exists. In real life, the biomedical sciences have not yet reached the stage of any kind of general applicability to disease mechanisms. In some respects we are like the physical sciences of the early twentieth century, booming along into new territory, but without an equivalent for the engineering of that time. It is possible that we are on the verge of developing a proper applied science, but it has to be said that we don't have one yet. The important question before the policy-makers is whether this should be allowed to occur naturally, as a matter of course, or whether it can be ordered up more quickly, under the influence of management and money.

There are risks. We may be asking for more of the kind of trouble with which we are already too familiar. There is a trap here that has enmeshed medicine for all the millennia of its professional existence. It has been our perpetual habit to try anything, on the slimmest of chances, the thinnest of hopes, empirically and wishfully, and we have proved to ourselves over and over again that the approach doesn't work well. Bleeding, cupping, and purging are the classical illustrations, but we have plenty of more recent examples to be embarrassed about. We have been hoaxed along by comparable substitutes for technology right up to the present. There is no question about our good intentions in this matter: we all hanker, collectively, to become applied scientists as soon as we can, overnight if possible.

It takes some doing, however. Everyone forgets how long and hard the work must be before the really important applications become applicable. The great contemporary achievement of modern medicine is the technology for controlling and preventing bacterial infection, but this did not fall into our laps with the appearance of penicillin and the sulfonamides. It had its beginnings in the final quarter of the last century, and decades of the most painstaking and demanding research were required before the etiology of pneumonia, scarlet fever, meningitis, and the rest could be worked out. Generations of energetic and imaginative investigators exhausted their whole lives on the problems. It overlooks a staggering amount of basic research to say that modern medicine began with the era of antibiotics.

We have to face, in whatever discomfort, the real possibility that the level of insight into the mechanisms of today's unsolved diseases—schizophrenia, for instance, or cancer, or stroke—is comparable to the situation for infectious disease in 1875, with similarly crucial bits of information still unencountered. We could be that far away, in the work to be done if not in the years to be lived through. If this is the prospect, or anything like this, all ideas about better ways to speed things up should be given open-minded, close scrutiny.

Long-range planning and organization on a national scale are obviously essential. There is nothing unfamiliar about this; indeed, we've been engaged in a coordinated national effort for over two decades, through the established processes of the National Institutes of Health. Today's question is whether the plans are sharply focused enough, the organization sufficiently tight. Do we need a new system of research management, with all the targets in clear display, arranged to be aimed at?

This would seem reassuring and tidy, and there are some

important disease problems for which it has already been done effectively, demonstrating that the direct, frontal approach does work. Poliomyelitis is the most spectacular example. Once it had been learned (from basic research) that there were three antigenic types of virus and that they could be abundantly grown in tissue culture, it became a certainty that a vaccine could be made. Not to say that the job would be easy, or in need of any less rigor and sophistication than the previous research; simply that it could be done. Given the assumption that experiments would be carried out with technical perfection, the vaccine was a sure thing. It was an elegant demonstration of how to organize applied science, and for this reason it would have been a surprise if it had not succeeded.

This is the element that distinguishes applied science from basic. Surprise is what makes the difference. When you are organized to apply knowledge, set up targets, produce a usable product, you require a high degree of certainty from the outset. All the facts on which you base protocols must be reasonably hard facts with unambiguous meaning. The challenge is to plan the work and organize the workers so that it will come out precisely as predicted. For this, you need centralized authority, elaborately detailed time schedules, and some sort of reward system based on speed and perfection. But most of all you need the intelligible basic facts to begin with, and these must come from basic research. There is no other source.

In basic research, everything is just the opposite. What you need at the outset is a high degree of uncertainty; otherwise it isn't likely to be an important problem. You start with an incomplete roster of facts, characterized by their ambiguity; often the problem consists of discovering the connections between unrelated pieces of information. You must plan experiments on the basis of probability, even bare possibility, rather than certainty. If an experiment turns out precisely as predicted, this can be very nice, but it is only a great event if at the same time it is a surprise. You can measure the quality of the work by the intensity of astonishment. The surprise can be because it did turn out as predicted (in some lines of research, 1 per cent is accepted as a high yield), or it can be confoundment because the prediction was wrong and something totally unexpected turned up, changing the look of the problem and requiring a new kind of protocol. Either way, you win.

I believe, on hunch, that an inventory of our major disease problems based on this sort of classification would show a limited number of important questions for which the predictable answers carry certainty. It might be a good idea, when commissions go to work laying out long-range plans for disease-oriented research, for these questions to be identified and segregated from all the rest, and the logic of operations research should be invaluable for this purpose. There will be lots of disputing among the experts over what is certain and what not; perhaps the heat and duration of dispute could be adapted for the measurement of uncertainty. In any case, once a set of suitable questions becomes agreed upon, these can be approached by the most systematic methods of applied science.

However, I have a stronger hunch that the greatest part of the important biomedical research waiting to be done is in the class of basic science. There is an abundance of interesting fact relating to all our major diseases, and more items of information are coming in steadily from all quarters in biology. The new mass of knowledge is still formless, incomplete, lacking the essential threads of connection, displaying misleading signals at every turn, riddled with blind alleys. There are fascinating ideas all over the place, irresistible experiments beyond numbering, all sorts of new ways into the maze of problems. But every next move is unpredictable, every outcome uncertain. It is a puzzling time, but a very good time.

I do not know how you lay out orderly plans for this kind

of activity, but I suppose you could find out by looking through the disorderly records of the past hundred years. Somehow, the atmosphere has to be set so that a disquieting sense of being wrong is the normal attitude of the investigators. It has to be taken for granted that the only way in is by riding the unencumbered human imagination, with the special rigor required for recognizing that something can be highly improbable, maybe almost impossible, and at the same time true.

Locally, a good way to tell how the work is going is to listen in the corridors. If you hear the word, "Impossible!" spoken as an expletive, followed by laughter, you will know that someone's orderly research plan is coming along nicely.

Some
Biomythology

The mythical animals catalogued in the bestiaries of the world seem, at a casual glance, nothing but exotic nonsense. The thought comes that Western civilized, scientific, technologic society is a standing proof of human progress, in having risen above such imaginings. They are as obsolete as the old anecdotes in which they played their puzzling, ambiguous roles, and we have no more need for the beasts than for the stories. The Griffon, Phoenix, Centaur, Sphinx, Manticore, Ganesha, Ch'i-lin, and all the rest are like recurrent bad dreams, and we are well rid of them. So we say.

The trouble is that they are in fact like dreams, and not necessarily bad ones, and we may have a hard time doing without them. They may be as essential for society as mythology itself, as loaded with symbols, and as necessary for the architecture of our collective unconscious. If Lévi-Strauss is right, myths are constructed by a universal logic that, like language itself, is as characteristic for human beings as nest-building is for birds. The stories seem to be different stories, but the underlying structure is always the same, in any part of the world, at any time. They are like engrams, built into our genes. In this sense, bestiaries are part of our inheritance.

There is something basically similar about most of these crazy animals. They are all unbiologic, but unbiologic in the same way. Bestiaries do not contain, as a rule, totally novel creatures of the imagination made up of parts that we have never seen before. On the contrary, they are made up of parts that are entirely familiar. What is novel, and startling, is that they are mixtures of species.

It is perhaps this characteristic that makes the usual bestiary so outlandish to the twentieth-century mind. Our most powerful story, equivalent in its way to a universal myth, is evolution. Never mind that it is true whereas myths are not; it is filled with symbolism, and this is the way it has influenced the mind of society. In our latest enlightenment, the fabulous beasts are worse than improbable—they are impossible, because they violate evolution. They are not species, and they deny the existence of species.

The Phoenix comes the closest to being a conventional animal, all bird for all of its adult life. It is, in fact, the most exuberant, elaborate, and ornamented of all plumed birds. It exists in the mythology of Egypt, Greece, the Middle East, and Europe, and is the same as the vermilion bird of ancient China. It lives for five hundred triumphant years, and when it dies it constructs a sort of egg-shaped cocoon around itself. Inside, it disintegrates and gives rise to a wormlike creature, which then develops into the new Phoenix, ready for the next five hundred years. In other versions the dead bird bursts into flames, and the new one arises from the ashes, but the worm story is very old, told no doubt by an early biologist.

There are so many examples of hybrid beings in bestiaries that you could say that an ardent belief in mixed forms of life is an ancient human idea, or that something else, deeply believed in, is symbolized by these consortia. They are disturbing to look at, nightmarish, but most of them, oddly enough, are intended as lucky benignities. The Ch'i-lin, for instance, out of ancient China, has the body of a deer covered with gleaming scales, a marvelous bushy tail, cloven hooves, and small horns. Whoever saw a Ch'i-lin was in luck, and if you got to ride one, you had it made.

The Ganesha is one of the oldest and most familiar Hindu deities, possessing a fat human body, four human arms, and the head of a cheerful-looking elephant. Prayers to Ganesha are regarded as the quickest way around obstacles.

Not all mythical beasts are friendly, of course, but even the hostile ones have certain amiable redeeming aspects. The Manticore has a lion's body, a man's face, and a tail with a venomous snake's head at the end of it. It bounds around seeking prey with huge claws and three rows of teeth, but it makes the sounds of a beautiful silver flute.

Some of the animal myths have the ring of contemporary biologic theory, if you allow for differences in jargon. An ancient idea in India postulates an initial Being, the first form of life on the earth, analogous to our version of the earliest prokaryotic arrangement of membrane-limited nucleic acid, the initial cell, born of lightning and methane. The Indian Being, undefined and indefinable, finding itself alone, fearing death, yearning for company, began to swell in size, rearranged itself inside, and then split into two identical halves. One of these changed into a cow, the other a bull, and they mated, then changed again to a mare and stallion, and so on, down to the ants, and thus the earth was populated. There is a lot of oversimplification here, and too much shorthand for modern purposes, but the essential myth is recognizable.

The serpent keeps recurring through the earliest cycles of mythology, always as a central symbol for the life of the universe and the continuity of creation. There are two great identical snakes on a Levantine libation vase of around 2000 B.C., coiled around each other in a double helix, representing the original generation of life. They are the replicated parts of the first source of living, and they are wonderfully homologous.

There is a Peruvian deity, painted on a clay pot dating

from around A.D. 300, believed to be responsible for guarding farms. His hair is made of snakes, entwined in braids, with wings for his headdress. Plants of various kinds are growing out of his sides and back, and a vegetable of some sort seems to be growing from his mouth. The whole effect is wild and disheveled but essentially friendly. He is, in fact, an imaginary version of a genuine animal, *symbiopholus,* described in *Nature* several years back, a species of weevil in the mountains of northern New Guinea that lives symbiotically with dozens of plants, growing in the niches and clefts in its carapace, rooted all the way down to its flesh, plus a whole ecosystem of mites, rotifers, nematodes, and bacteria attached to the garden. The weevil could be taken for a good-luck omen on its own evidence; it is not attacked by predators, it lives a long, untroubled life, and nothing else will eat it, either because of something distasteful in the system or simply because of the ambiguity. The weevil is only about thirty millimeters long, easily overlooked, but it has the makings of a myth.

Perhaps we should be looking around for other candidates. I suggest the need for a new bestiary, to take the place of the old ones. I can think of several creatures that seem designed for this function, if you will accept a microbestiary, and if you are looking for metaphors.

First of all, there is *Myxotricha paradoxa.* This is the protozoan, not yet as famous as he should be, who seems to be telling us everything about everything, all at once. His cilia are not cilia at all, but individual spirochetes, and at the base of attachment of each spirochete is an oval organelle, embedded in the myxotricha membrane, which is a bacterium. It is not an animal after all—it is a company, an assemblage.

The story told by myxotricha is as deep as any myth, as profoundly allusive. This creature has lagged behind the rest of us, and is still going through the process of being assembled. Our cilia gave up any independent existence long ago, and our organelles are now truly ours, but the genomes controlling separate parts of our cells are still different genomes, lodged in separate compartments; doctrinally, we are still assemblages.

There is another protozoan, called blepharisma, telling a long story about the chanciness and fallibility of complex life. Blepharisma is called that because of a conspicuous fringe of ciliated membranes around the oral cavity, which evidently reminded someone of eyelashes *(blepharidos)*. The whole mythlike tale has been related in a book by Giese. Blepharisma has come much further along than myxotricha, but not far enough to be free of slip-ups. There are three different sets of self-duplicating nuclei, with the DNA in each set serving different purposes: a large macronucleus, governing the events in regeneration after injury, a set of eight or more micronuclei containing the parts of the genome needed for reproduction, and great numbers of tiny nuclei from which the cilia arise.

One part of the organism produces a pinkish pigment, now called blepharismin, which is similar to hypericin and certain other photosensitizing plant pigments. Blepharismin causes no trouble unless the animal swims into sunlight, but then the pigment kills it outright. Under certain circumstances, the membrane surrounding blepharisma disintegrates and comes independently loose, like a cast-off shell, leaving the creature a transient albino. At times of famine, a single blepharisma will begin eating its neighbors; it then enlarges to an immense size and turns into a cannibalistic giant, straight out of any Norse fable. Evidently, this creature still has trouble getting along with the several parts of itself, and with the collective parts of other blepharismae.

There are innumerable plant-animal combinations, mostly in the sea, where the green plant cells provide carbohydrate and oxygen for the animal and receive a share of energy in return. It is the fairest of arrangements. When the paramecium bursaria runs out of food, all he needs to do is stay in the sun and his green endosymbionts will keep him supplied as though he were a grain.

Bacteria are the greatest of all at setting up joint enterprises, on which the lives of their hosts are totally dependent. The nitrogen-fixing rhizobia in root nodules, the mycetomes of insects, and the enzyme-producing colonies in the digestive tracts of many animals are variations of this meticulously symmetrical symbiosis.

The meaning of these stories may be basically the same as the meaning of a medieval bestiary. There is a tendency for living things to join up, establish linkages, live inside each other, return to earlier arrangements, get along, whenever possible. This is the way of the world.

The new phenomenon of cell fusion, a laboratory trick on which much of today's science of molecular genetics relies for its data, is the simplest and most spectacular symbol of the tendency. In a way, it is the most unbiologic of all phenomena, violating the most fundamental myth of the last century, for it denies the importance of specificity, integrity, and separateness in living things. Any cell—man, animal, fish, fowl, or insect—given the chance and under the right conditions, brought into contact with any other cell, however foreign, will fuse with it. Cytoplasm will flow easily from one to the other, the nuclei will combine, and it will become, for a time anyway, a single cell with two complete, alien genomes, ready to dance, ready to multiply. It is a Chimera, a Griffon, a Sphinx, a Ganesha, a Peruvian god, a Ch'i-lin, an omen of good fortune, a wish for the world.

On Various Words

T he idea that colonies of social insects are somehow equivalent to vast, multicreatured organisms, possessing a collective intelligence and a gift for adaptation far superior to the sum of the individual inhabitants, had its origin in the papers of the eminent entomologist, William Morton Wheeler, who proposed the term Superorganism to describe the arrangement. From 1911 to the early 1950s this ranked as a central notion in entomology, attracting the attention of many fascinated nonentomologists. Maeterlinck and Marais wrote best-selling books on the presumed soul that must exist somewhere in the nests of ants and termites.

Then, unaccountably, the whole idea abruptly dropped out of fashion and sight. During the past quarter-century almost no mention of it is made in the proliferation of scientific literature in entomology. It is not talked about. It is not just that the idea has been forgotten; it is as though it had become unmentionable, an embarrassment.

It is hard to explain. The notion was not shown to be all that mistaken, nor was it in conflict with any other, more acceptable view of things. It was simply that nobody could figure out what to do with such an abstraction. There it sat, occupying important intellectual ground, at just the time when entomology was emerging as an experimental science of considerable power, capable of solving matters of intricate detail, a paradigm of the new reductionism. This huge idea—that individual organisms might be self-transcending in their relation to a dense society—was not approachable by the new techniques, nor did it suggest new experiments or methods. It just sat there, in the way, and was covered over by leaves and papers. It needed heuristic value to survive, and this was lacking.

"Holism," a fabricated word, has been applied to concepts like the Superorganism. One wonders whether this word may not itself have scared off some investigators; it is a word with an alarming visage. General Jan Smuts, who invented it out of whole cloth in 1926, might have done better with "Wholism"; it would have served the same etymological purpose and might have been just secular enough to survive this kind of century. As it is, there is doubt for its future. Holism is in some of the scientific glossaries but has not yet made it into most standard dictionaries of English. It got as far as the Supplement volume of the new OED, which is something, but not enough to assure survival. Perhaps it will die away, along with Superorganism.

I cannot quarrel with any of this. If an idea cannot move on its own, pushing it doesn't help; best to let it lie there.

It may be, though, that the pushing was tried in the wrong direction. Colonies of ants or termites, or bees and social wasps, may in fact be Superorganisms by Wheeler's criteria, but perhaps that is the end of that line of information as far as insects are concerned, for the time being. Maybe it would work better if you tried it out on another social species, easier to handle. Us, for one.

It has long troubled the entomologists that the rest of us are always interfering in their affairs by offering explanations of insect behavior in human terms. They take pains to explain that ants are not, emphatically not, tiny mechanical models of human beings. I agree with this. Nothing that we know for sure about human behavior is likely to account for what ants do, and we ought to stay clear of it; this is the business of entomologists. As for the ants themselves, they are plainly not in need of lessons from us.

However, this does not mean that we cannot take it the other way, on the off chance that some of the collective actions of ants may cast light on human problems.

There are lots of possibilities here, but if you think about the construction of the Hill by a colony of a million ants, each one working ceaselessly and compulsively to add perfection to his region of the structure without having the faintest notion of what is being constructed elsewhere, living out his brief life in a social enterprise that extends back into what is for him the deepest antiquity (ants die at the rate of 3–4 per cent per day; in a month or so an entire generation vanishes, while the Hill can go on for sixty years or, given good years, forever), performing his work with infallible, undistracted skill in the midst of a confusion of others, all tumbling over each other to get the twigs and bits of earth aligned in precisely the right configurations for the warmth and ventilation of the eggs and larvae, but totally incapacitated by isolation, there is only one human activity that is like this, and it is language.

We have been working at it for what seems eternity, generation after articulate generation, and still we have no notion how it is done, nor what it will be like when finished, if it is ever to be finished. It is the most compulsively collective, genetically programmed, species-specific, and autonomic of all the things we do, and we are infallible at it. It comes naturally. We have DNA for grammar, neurons for syntax. We can never let up; we scramble our way through one civilization after another, metamorphosing, sprouting tools and cities everywhere, and all the time new words keep tumbling out.

The words themselves are marvels, each one perfectly designed for its use. The older, more powerful ones are membranous, packed with layers of different meaning, like one-word poems. "Articulated," for instance, first indicated a division into small joints, then, effortlessly, signified the speaking of sentences. Some words are gradually altered while we have them in everyday use, without our being aware until the change has been completed: the *ly* in today's adverbs, such as ably and benignly, began to appear in place of "like" just a few centuries ago, and "like" has since worn away to a mere suffix. By a similar process, "love-did" changed itself into "loved."

None of the words are ever made up by anyone we know; they simply turn up in the language when they are needed. Sometimes a familiar word will suddenly be grabbed up and transformed to mean something quite strange: "strange" is itself such a word today, needed by nuclear physicists to symbolize the behavior of particles which decay with peculiar slowness; the technical term for such particles now is "strange particles," and they possess a "strangeness number(S)." The shock of sudden unfamiliarity with an old, familiar word is something we take in stride; it has been going on for thousands of years.

A few words are made up by solitary men in front of our eyes, like Holism out of Smuts, or Quark out of Joyce, but most of these are exotic and transient; it takes a great deal of use before a word can become a word.

Most new words are made up from other, earlier words; language-making is a conservative process, wasting little. When new words unfold out of old ones, the original meaning usually hangs around like an unrecognizable scent, a sort of secret.

"Holism" suggests something biologically transcendental because of "holy," although it was intended more simply to mean a complete assemblage of living units. Originally, it came from the Indo-European root word *kailo,* which meant whole, also intact and uninjured. During passage through several thousand years it transformed into hail, hale, health, hallow, holy, whole, and heal, and all of these still move together through our minds.

"Heuristic" is a more specialized, single-purpose word, derived from Indo-European *wer,* meaning to find, then taken up in Greek as *heuriskein,* from which Archimedes was provided with Heureka!

There are two immense words from Indo-European, *gene* and *bheu,* each a virtual anthill in itself, from which we have constructed the notion of Everything. At the beginning, or as far back as they are traceable, they meant something like being. *Gene* signified beginning, giving birth, while *bheu* indicated existence and growth. *Gene* turned itself successively into *kundjaz* (Germanic) and *gecynd* (Old English), meaning kin or kind. Kind was at first a family connection, later an elevated social rank, and finally came to rest meaning kindly or gentle. Meanwhile, a branch of *gene* became the Latin *gens,* then gentle itself; it also emerged as genus, genius, genital, and generous; then, still holding on to its inner significance, it became "nature" (out of *gnasci*).

While gene was evolving into "nature" and "kind," *bheu* was moving through similar transformations. One branch became *bowan* in Germanic and *bua* in Old Norse, meaning to live and dwell, and then the English word build. It moved into Greek, as *phuein,* meaning to bring forth and make grow; then as *phusis,* which was another word for nature. *Phusis* became the source of physic, which at first meant natural science and later was the word for medicine. Still later, physic became physics.

Both words, at today's stage of their evolution, can be taken together to mean, literally, everything in the universe. You do not come by words like this easily; they cannot just be made up from scratch. They need long lives before they can signify. "Everything," C.S. Lewis observed in a discussion of the words, "is a subject on which there is not much to be said." The words themselves must show the internal marks of long use; they must contain their own inner conversation.

These days it is reassuring to know that nature and physics, in their present meanings, have been interconnected in our minds, by a sort of hunch, for all these years. The other words clinging to them are a puzzlement, but nice to see. If you let your mind relax, all the words will flow into each other in an amiable sort of nonsense. "Kind" means a relation, but it also means "nature." The word for kind is the same as the word for gentle. Even "physics," save us, is a kind of "nature," by its nature, and is, simultaneously, another kind of kind. There are ancient ideas reverberating through this structure, very old hunches.

It is part of the magic of language that some people can get to the same place by the use of totally different words. Julian of Norwich, a fourteenth-century hermitess, said it so well that a paragraph of hers was used recently by a physicist for his introduction to a hard-science review of contemporary cosmological physics: "He showed me a little thing, the quantity of an hazelnut, in the palm of my hand, and it was as round as a ball. I looked thereupon with eye of my understanding and thought: What may this be? And it was answered generally thus: it is all that is made."

Living
Language

"Stigmergy" is a new word, invented recently by Grassé to explain the nest-building behavior of termites, perhaps generalizable to other complex activities of social animals. The word is made of Greek roots meaning "to incite to work," and Grassé's intention was to indicate that it is the product of work itself that provides both the stimulus and instructions for further work. He arrived at this after long observation of the construction of termite nests, which excepting perhaps a man-made city are the most formidable edifices in nature. When you consider the size of an individual termite, photographed standing alongside his nest, he ranks with the New Yorker and shows a better sense of organization than a resident of Los Angeles. Some of the mound nests of *Macrotermes bellicosus* in Africa measure twelve feet high and a hundred feet across, they contain several millions of termites, and around them are clustered other small and younger mounds, like suburbs.

The interior of the nests are like a three-dimensional maze, intricate arrangements of spiraling galleries, corridors, and arched vaults, ventilated and air-conditioned. There are great caverns for the gardens of fungi on which the termites depend for their nourishment, perhaps also as a source of heat. There is a rounded vaulted chamber for the queen, called the royal cell. The fundamental structural unit, on which the whole design is based, is the arch.

Grassé needed his word in order to account for the ability of such tiny, blind, and relatively brainless animals to erect structures of such vast size and internal complexity. Does each termite possess a fragment of blueprint, or is the whole design, arch by arch, encoded in his DNA? Or does the whole colony have, by virtue of the interconnections of so many small brains, the collective intellectual power of a huge contractor?

Grassé placed a handful of termites in a dish filled with soil and fecal pellets (these are made of lignin, a sort of micro-lumber) and watched what they did. They did not, in the first place, behave at all like contractors. Nobody stood around in place and gave orders or collected fees; they all simply ran around, picking up pellets at random and dropping them again. Then, by chance, two or three pellets happened to light on top of each other, and this transformed the behavior of everyone. Now they displayed the greatest interest and directed their attention obsessively to the primitive column, adding new pellets and fragments of earth. After reaching a certain height, the construction stopped unless another column was being formed nearby; in this case the structure changed from a column to an arch, bending off in a smooth curve, the arch was joined, and the termites then set off to build another.

Building a language may be something like this. One can imagine primitive proto-Indo-European men finding themselves clustered together, making random sounds, surrounded, say, by bees, and one of them suddenly saying *"bhei,"* and then the rest of them picking it up and repeating *"bhei"* and thus beginning that part of language, but this is a restricted, too mechanistic view of things. It makes pellets out of phonemes, implies that the deep structures of grammar are made of something like cement. I do not care for this.

More likely, language is simply alive, like an organism.

We tell each other this, in fact, when we speak of living languages, and I think we mean something more than an abstract metaphor. We mean alive. Words are the cells of language, moving the great body, on legs.

Language grows and evolves, leaving fossils behind. The individual words are like different species of animals. Mutations occur. Words fuse, and then mate. Hybrid words and wild varieties of compound words are the progeny. Some mixed words are dominated by one parent while the other is recessive. The way a word is used this year is its phenotype, but it has a deeply seated, immutable meaning, often hidden, which is the genotype.

The language of genetics might be used in some such way to describe the genetics of language, if we knew more about both.

The separate languages of the Indo-European family were at one time, perhaps five thousand years ago, maybe much longer, a single language. The separation of the speakers by migrations had effects on language comparable to the speciation observed by Darwin on various islands of Galapagos. Languages became different species, retaining enough resemblance to an original ancestor so that the family resemblance can still be seen. Variation has been maintained by occasional contact between different islands of speakers, and perhaps also by random mutations.

But there is something else about words that gives them the look and feel of living motile beings with minds of their own. This is best experienced by looking them up, preferably in one of the dictionaries that provide all the roots back to the original, hypothetical fossil language of proto-Indo-European, and observing their behavior.

Some words started from Indo-European and swarmed into religion over a very large part of the earth. The word *blaghmen,* for example, meant priest. It moved into Latin and Middle English as *flamen,* a pagan word for priest, and into Sanskrit as *brahma,* then "brahman." *Weid,* a word meaning to see, with later connotations of wisdom and wit, entered Germanic as *witan,* and Old English *wis* to "wisdom." It became *videre* in Latin, hence "vision." Finally, in its suffixed form *woid-o,* it became the Sanskrit word *veda.*

Beudh traveled a similar distance. With the meaning of awareness, it became *beodan* in Old English, meaning "bode," and *bodhati* in Sanskrit, meaning he awakes, is enlightened, and thus Bodhisattva, and Buddha.

The *sattva* part of Bodhisattva came from the Indo-European word *es,* meaning to be, or is, which, on its way into Sanskrit as *sat* and *sant,* also became *esse* in Latin and *einai* in Greek; einai became the -ont in certain words signifying being, such as "symbiont."

The Indo-European word *bhag,* meaning to share, turned into the Greek *phagein,* to eat, and the Old Persian *bakhsh* (yielding "baksheesh," and, in Sanskrit, with the meaning of *bhage,* good fortune, it emerged as Bhagavad-gita (the *gita* from *gei,* a song).

The Hari-Krishna people are chanting something closer to English than it sounds. Krishna, the eighth avatar of Vishnu, has his name from the Sanskrit *Krsnah,* the black one, which came from the Indo-European word for black, *kers* (which also produced "chernozem," black topsoil, by the way of the Russian *chernyi*).

There is obviously no end to this; it can tie up a whole life, and has luckily done just that during the past century for generations of comparative linguists. Their science began properly in 1786 with the discovery of the similarity of Sanskrit to Greek and Latin, by William Jones. In 1817, with a publication by Franz Bopp, it became recognized that Sanskrit, Greek, Latin, Persian, and all the Germanic languages were so closely related to each other that a common ancestor must have existed earlier. Since then, this science has developed in more or less parallel with biology, but more quietly.

It is a field in which the irresponsible amateur can have a continually mystifying sort of fun. Whenever you get the available answer to a straight question, like, say, where does the most famous and worst of the four-letter Anglo-Saxon unprintable words come from, the answer raises new and discomfiting questions. Take that particular word. It comes from *peig,* a crawling, wicked Indo-European word meaning evil and hostile, the sure makings of a curse. It becomes *poikos,* then *gafaihaz* in Germanic and *gefah* in Old English, signifying "foe." It turned from *poik-yos* into *faigjaz* in Germanic, and *faege* in Old English, meaning fated to die, leading to "fey." It went on from *fehida* in Old English to become "feud," and *fokken* in Old Dutch. Somehow, from these beginnings, it transformed itself into one of the most powerful English expletives, meaning something like "Die before your time!" The unspeakable malevolence of the message is now buried deep inside the word, and out on the surface it presents itself as merely an obscenity.

"Leech" is a fascinating word. It is an antique term for physician, and also for the aquatic worm *sanguisugus,* used for leeching. The two words appear to be quite separate, but there is something like biological mimicry going on: leech the doctor means the doctor who uses leech the worm; leech the worm is a symbol for the doctor. Leech the doctor comes from the Indo-European *leg,* which meant to collect, with numerous derivatives meaning to speak. *Leg* became Germanic *lekjaz,* meaning one who speaks magic words, an enchanter, and also *laece* in Old English, meaning physician. (In Denmark the word for doctor is still *laege,* in Swedish *läkare.*) *Leg* in its senses of gathering, choosing, and speaking gave rise to the Latin *legere,* and thus words like "lecture" and "legible." In Greek, it became *legein,* meaning to gather and to speak; "legal" and "legislator" and other such words derived. *Leg* was further transformed in Greek to *logos,* signifying reason.

All this history seems both plausible and creditable, good reading for doctors, but there is always that other leech, the worm. It is not certain how it came. Somehow it began its descent through the language at the same time as leech the doctor, turning up as both *laece* and *lyce* in Old English, always recognizable as something distinctly the worm and at the same time important in medicine. It also took on the meaning of someone parasitic, living on the flesh of others. Gradually, perhaps under the influence of a Middle English AMA, the worm was given sole rights to the word, and the doctor became the doctor, out of *dek,* meaning to accept, later to teach.

Man is an unchanged word from Indo-European *man,* meaning just that. But two other important words for man have stranger sources. One is *dhghem,* meaning earth; this became *guman* in Germanic, *gumen* in Old English, then *homo* and *humanus* in Latin, from which we have both "human" and "humus." The other word for man contains the same admonition, but turns the message around. It is *wiros,* meaning man in Indo-European, taken as *weraldh* in Germanic and *weorold* in Old English, emerging, flabbergastingly, as "world."

This must be a hard science to work in. You might think that with a word for earth giving rise to one important word for man, and an early word for man turning into the word for the world, you would find a parallel development in other words for the earth. Not so: the Indo-European word *ers,* which later became "earth," has evolved only one animal that I can find mentioned, and it is the aardvark.

I am glad to have a semipermeable memory after getting into this. If you had to speak English with running captions in your mind showing all the roots, all the way back to Indo-European, you'd fall off the bicycle. Speaking is an autonomic business; you may search for words as you go along, but they are found for you by agents in your brain over which you exercise no direct control. You really couldn't be thinking Indo-European at the same time, without going speechless or babbling (from *baba,* meaning indistinct speech, Russian *balalayka,* Latin *balbus,* meaning "booby," Old French *baboue,* leading to "baboon," Greek *barbaros,* meaning foreign or rude, and Sanskrit *babu,* meaning father). That sort of thing.

I got into even more trouble while looking into "Stigmergy." I was looking for other words for inciting and instigating work, and came upon "to egg on." The egg here comes from *ak,* a word for sharp, suffixed to *akjo* in Germanic, meaning "edge," and to *akjan* in Old Norse, meaning "egg," to incite, goad; the same root moves on into Old English as *aehher* and *ear,* for ear of corn. (Corn, if you have a moment, is from *greno,* for grain, which became *korn* in Old High German, *granum* in Latin, and *cyrnel* in Old English, thence "kernel.") But neither the egg nor the ear from *ak* are the real egg or ear. The real egg comes from *awi,* meaning bird, which turned into *avis* and *ovum* in Latin (not known, of course, which came first), into *oion* in Greek, and was compounded with *spek* (to see) to form *awispek,* "watcher of birds" which became *auspex* in Latin, meaning augur.

The real ear began as *ous,* then *auzan* in Germanic, and *eare* in Old English and *auri* in Latin; along the way, it was compounded with *sleg,* meaning slack, and transformed to *lagous,* meaning "with drooping ears," which then became *lagos,* Greek for rabbit.

There is no way to stop, once you've started, not even by trying to round a circle. *Ous* became *aus* became "auscultation," which is what leeches *(leg)* do for a living *(leip)* unless they are legal *(leg)* leeches, which, incidentally, is not the same thing as lawyers *(legh).*

That should be enough (*nek,* to attain, becoming *ganoga* in Germanic and *genog* in Old English, also *onkos* in Greek, meaning burden, hence "oncology") to give you the general *(gene)* idea (*weid* becoming *widesya* then *idea* in Greek). It is easy to lose the thread (from *ter,* to rub, twist—possibly also the root of termite). Are you there?

On Probability
and Possibility

Statistically, the probability of any one of us being here is so small that you'd think the mere fact of existing would keep us all in a contented dazzlement of surprise. We are alive against the stupendous odds of genetics, infinitely outnumbered by all the alternates who might, except for luck, be in our places.

Even more astounding is our statistical improbability in physical terms. The normal, predictable state of matter throughout the universe is randomness, a relaxed sort of equilibrium, with atoms and their particles scattered around in an amorphous muddle. We, in brilliant contrast, are completely organized structures, squirming with information at every covalent bond. We make our living by catching electrons at the moment of their excitement by solar photons, swiping the energy released at the instant of each jump and storing it up in intricate loops for ourselves. We violate probability, by our nature. To be able to do this systemically, and in such wild varieties of form, from viruses to whales, is extremely unlikely; to have sustained the effort successfully for the several billion years of our existence, without drifting back into randomness, was nearly a mathematical impossibility.

Add to this the biological improbability that makes each member of our own species unique. Everyone is one in 3 billion at the moment, which describes the odds. Each of us is a self-contained, free-standing individual, labeled by spe-

141

cific protein configurations at the surfaces of cells, identifiable by whorls of fingertip skin, maybe even by special medleys of fragrance. You'd think we'd never stop dancing.

Perhaps it is not surprising that we do not live more surprised. After all, we are used to unlikelihood. Being born into it, raised in it, we become acclimated to the altitude, like natives in the Andes. Moreover, we all know that the astonishment is transient, and sooner or later our particles will all go back to being random.

Also, there are reasons to suspect that we are really not the absolute, pure entities that we seem. We have some sense of ordinariness, and it tends to diminish our surprise. Despite all the evidences of biological privacy in our cells and tissues (to the extent that a fragment of cell membrane will be recognized and rejected between any conceivable pairs among the 3 billion, excepting identical twins), there is a certain slippage in our brains. No one, in fact, can lay claim with certainty to his own mind with anything like the specificity stipulated by fingerprints or tissue antigens.

The human brain is the most public organ on the face of the earth, open to everything, sending out messages to everything. To be sure, it is hidden away in bone and conducts internal affairs in secrecy, but virtually all the business is the direct result of thinking that has already occurred in other minds. We pass thoughts around, from mind to mind, so compulsively and with such speed that the brains of mankind often appear, functionally, to be undergoing fusion.

This is, when you think about it, really amazing. The whole dear notion of one's own Self—marvelous old free-willed, free-enterprising, autonomous, independent, isolated island of a Self—is a myth.

We do not yet have a science strong enough to displace the myth. If you could label, by some equivalent of radioactive isotopes, all the bits of human thought that are constantly adrift, like plankton, all around us, it might be possible to discern some sort of systematic order in the process, but, as it is, it seems almost entirely random. There has to be something wrong with this view. It is hard to see how we could be in possession of an organ so complex and intricate and, as it occasionally reveals itself, so powerful, and be using it on such a scale just for the production of a kind of background noise. Somewhere, obscured by the snatches of conversation, pages of old letters, bits of books and magazines, memories of old movies, and the disorder of radio and television, there ought to be more intelligible signals.

Or perhaps we are only at the beginning of learning to use the system, with almost all our evolution as a species still ahead of us. Maybe the thoughts we generate today and flick around from mind to mind, like the jokes that turn up simultaneously at dinner parties in Hong Kong and Boston, or the sudden changes in the way we wear our hair, or all the popular love songs, are the primitive precursors of more complicated, polymerized structures that will come later, analogous to the prokaryotic cells that drifted through shallow pools in the early days of biological evolution. Later, when the time is right, there may be fusion and symbiosis among the bits, and then we will see eukaryotic thought, metazoans of thought, huge interliving coral shoals of thought.

The mechanism is there, and there is no doubt that it is already capable of functioning, even though the total yield thus far seems to consist largely of bits. After all, it has to be said that we've been at it for only the briefest time in evolutionary terms, a few thousand years out of billions, and during most of this time the scattered aggregates of human thought have been located patchily around the earth. There may be some laws about this kind of communication, mandating a critical density and mass before it can function with efficiency. Only in this century have we been brought close enough to each other, in great numbers, to begin the fusion around the earth, and from now on the process may move very rapidly.

There is, if it goes well, quite a lot to look forward to. Already, by luck, we have seen the assembly of particles of exchanged thought into today's structures of art and science. It is done by simply passing the bits around from mind to mind, until something like natural selection makes the final selection, all on grounds of fitness.

The real surprises, which set us back on our heels when they occur, will always be the mutants. We have already had a few of these, sweeping across the field of human thought periodically, like comets. They have slightly different receptors for the information cascading in from other minds, and slightly different machinery for processing it, so that what comes out to rejoin the flow is novel, and filled with new sorts of meaning. Bach was able to do this, and what emerged in the current were primordia in music. In this sense, the Art of Fugue and the St. Matthew Passion were, for the evolving organism of human thought, feathered wings, apposing thumbs, new layers of frontal cortex.

But we may not be so dependent on mutants from here on, or perhaps there are more of them around than we recognize. What we need is more crowding, more unrestrained and obsessive communication, more open channels, even more noise, and a bit more luck. We are simultaneously participants and bystanders, which is a puzzling role to play. As participants, we have no choice in the matter; this is what we do as a species. As bystanders, stand back and give it room is my advice.

The World's
Biggest Membrane

Viewed from the distance of the moon, the astonishing thing about the earth, catching the breath, is that it is alive. The photographs show the dry, pounded surface of the moon in the foreground, dead as an old bone. Aloft, floating free beneath the moist, gleaming membrane of bright blue sky, is the rising earth, the only exuberant thing in this part of the cosmos. If you could look long enough, you would see the swirling of the great drifts of white cloud, covering and uncovering the half-hidden masses of land. If you had been looking for a very long, geologic time, you could have seen the continents themselves in motion, drifting apart on their crustal plates, held afloat by the fire beneath. It has the organized, self-contained look of a live creature, full of information, marvelously skilled in handling the sun.

It takes a membrane to make sense out of disorder in biology. You have to be able to catch energy and hold it, storing precisely the needed amount and releasing it in measured shares. A cell does this, and so do the organelles inside. Each assemblage is poised in the flow of solar energy, tapping off energy from metabolic surrogates of the sun. To stay alive, you have to be able to hold out against equilibrium, maintain imbalance, bank against entropy, and you can only transact this business with membranes in our kind of world.

When the earth came alive it began constructing its own membrane, for the general purpose of editing the sun.

Originally, in the time of prebiotic elaboration of peptides and nucleotides from inorganic ingredients in the water on the earth, there was nothing to shield out ultraviolet radiation except the water itself. The first thin atmosphere came entirely from the degassing of the earth as it cooled, and there was only a vanishingly small trace of oxygen in it. Theoretically, there could have been some production of oxygen by photodissociation of water vapor in ultraviolet light, but not much. This process would have been self-limiting, as Urey showed, since the wave lengths needed for photolysis are the very ones screened out selectively by oxygen; the production of oxygen would have been cut off almost as soon as it occurred.

The formation of oxygen had to await the emergence of photosynthetic cells, and these were required to live in an environment with sufficient visible light for photosynthesis but shielded at the same time against lethal ultraviolet. Berkner and Marshall calculate that the green cells must therefore have been about ten meters below the surface of water, probably in pools and ponds shallow enough to lack strong convection currents (the ocean could not have been the starting place).

You could say that the breathing of oxygen into the atmosphere was the result of evolution, or you could turn it around and say that evolution was the result of oxygen. You can have it either way. Once the photosynthetic cells had appeared, very probably counterparts of today's blue-green algae, the future respiratory mechanism of the earth was set in place. Early on, when the level of oxygen had built up to around 1 per cent of today's atmospheric concentration, the anaerobic life of the earth was placed in jeopardy, and the inevitable next stage was the emergence of mutants with oxidative systems and ATP. With this, we were off to an explosive developmental stage in which great varieties of respiring life, including the multicellular forms, became feasible.

Berkner has suggested that there were two such explosions of new life, like vast embryological transformations, both dependent on threshold levels of oxygen. The first, at 1 per cent of the present level, shielded out enough ultraviolet radiation to permit cells to move into the surface layers of lakes, rivers, and oceans. This happened around 600 million years ago, at the beginning of the Paleozoic era, and accounts for the sudden abundance of marine fossils of all kinds in the record of this period. The second burst occurred when oxygen rose to 10 per cent of the present level. At this time, around 400 million years ago, there was a sufficient canopy to allow life out of the water and onto the land. From here on it was clear going, with nothing to restrain the variety of life except the limits of biologic inventiveness.

It is another illustration of our fantastic luck that oxygen filters out the very bands of ultraviolet light that are most devastating for nucleic acids and proteins, while allowing full penetration of the visible light needed for photosynthesis. If it had not been for this semipermeability, we could never have come along.

The earth breathes, in a certain sense. Berkner suggests that there may have been cycles of oxygen production and carbon dioxide consumption, depending on relative abundances of plant and animal life, with the ice ages representing periods of apnea. An overwhelming richness of vegetation may have caused the level of oxygen to rise above today's concentration, with a corresponding depletion of carbon dioxide. Such a drop in carbon dioxide may have impaired the "greenhouse" property of the atmosphere, which holds in the solar heat otherwise lost by radiation from the earth's surface. The fall in temperature would in turn have shut off much of living, and, in a long sigh, the level of oxygen may have dropped by 90 per cent. Berkner speculates that this is what happened to the great reptiles; their size may have been all right for a richly oxygenated atmosphere, but they had the bad luck to run out of air.

Now we are protected against lethal ultraviolet rays by a narrow rim of ozone, thirty miles out. We are safe, well ventilated, and incubated, provided we can avoid technologies that might fiddle with that ozone, or shift the levels of carbon dioxide. Oxygen is not a major worry for us, unless we let fly with enough nuclear explosives to kill off the green cells in the sea; if we do that, of course, we are in for strangling.

It is hard to feel affection for something as totally impersonal as the atmosphere, and yet there it is, as much a part and product of life as wine or bread. Taken all in all, the sky is a miraculous achievement. It works, and for what it is designed to accomplish it is as infallible as anything in nature. I doubt whether any of us could think of a way to improve on it, beyond maybe shifting a local cloud from here to there on occasion. The word "chance" does not serve to account well for structures of such magnificence. There may have been elements of luck in the emergence of chloroplasts, but once these things were on the scene, the evolution of the sky became absolutely ordained. Chance suggests alternatives, other possibilities, different solutions. This may be true for gills and swim-bladders and forebrains, matters of detail, but not for the sky. There was simply no other way to go.

We should credit it for what it is: for sheer size and perfection of function, it is far and away the grandest product of collaboration in all of nature.

It breathes for us, and it does another thing for our pleasure. Each day, millions of meteorites fall against the outer limits of the membrane and are burned to nothing by the friction. Without this shelter, our surface would long since have become the pounded powder of the moon. Even though our receptors are not sensitive enough to hear it, there is comfort in knowing that the sound is there overhead, like the random noise of rain on the roof at night.

Reference Notes

Thoughts for a Countdown

Hanks, J. H., "Host-Dependent Microbes," *Bacteriological Review*, 30:114–35, 1966.

Shilo, M., "Morphological and Physiological Aspects of the Interaction of *Bdellovibrio* with Host Bacteria," *Current Topics in Microbiology and Immunology*, 50:174–204, 1969.

Dilworth, M. J., "The Plant as the Genetic Determinant of Leghaemoglobin Production in the Legume Root Nodule," *Biochemica et Biophysica Acta*, 184:432–41, 1969.

Timourian, H., "Symbiotic Emergence of Metazoans," *Nature*, 226:283–84, 1970.

Gotto, R. V. *Marine Animals: Partnerships and Other Associations*. New York: American Elsevier, 1969.

Thompson, T. E., and Bennett, I., "Physalia Nematocysts: Utilized by Mollusks for Defense," *Science*, 166:1532–33, 1969.

Theodor, J. L., "The Distinction between 'Self' and 'Non-Self' in Lower Invertebrates," *Nature*, 227:690–692, 1970.

Parker, B. C., "Rain as a Source of Vitamin B_{12}," *Nature*, 219:617–18, 1968.

On Societies as Organisms

Ziman, J. M., "Information, Communication, Knowledge," *Nature*, 224:318–24, 1969.

A Fear of Pheromones

Comfort, A., "The Likelihood of Human Pheromones," *Nature,* 230:432–33, 1971.

Hoyt, C. P., Osborne, G. O., and Mulcock, A. P., "Production of an Insect Sex Attractant by Symbiotic Bacteria," *Nature,* 230:472–73, 1971.

Wilson, E. O., "Chemical Systems," in T. A. Seboek, ed., *Animal Communication: Techniques of Study and Results of Research.* Bloomington: Indiana University Press, 1970.

Todd, J. H., "The Chemical Languages of Fishes," *Scientific American,* 224(5):98–108, 1971.

Michael, R. P., Keverne, E. B., and Bonsall, R. W., "Pheromones: Isolation of Male Sex Attractants from a Female Primate," *Science,* 172:964–66, 1971.

McClintock, M. K., "Menstrual Synchrony and Suppression," *Nature,* 229:244–45, 1971.

"Effects of Sexual Activity on Beard Growth in Man," *Nature,* 226:869–70, 1970.

Smith, K., Thompson, G. F., and Koster, H. D., "Sweat in Schizophrenic Patients: Identification of the Odorous Substance," *Science,* 166:398–99, 1969.

The Music of This *Sphere*

Howse, P. E., "The Significance of the Sound Produced by the Termite *Zootermopsis angusticollis* (Hagen)," *Animal Behavior,* 12:284–300, 1964.

Busnel, R. G., ed., *Acoustic Behavior of Animals.* Amsterdam: Elsevier, 1963.

Payne, R. S., and McVay, S., "Songs of Humpback Whales," *Science* 173:585–97, 1971.

Morowitz, H. J., *Energy Flow in Biology: Biological Organization as a Problem in Thermal Physics.* New York: Academic Press, 1968.

An Earnest Proposal

Margulis, L., *The Origin of Eukaryotic Cells.* New Haven: Yale University Press, 1970.

Vibes

King, J. E., Becker, R. F., and Markee, J. E., "Studies on Olfactory Discrimination in Dogs," *Animal Behavior,* 12:-311–15, 1964.

Kalmus, H., "The Discrimination by the Nose of the Dog of Individual Human Odours and in Particular of the Odours of Twins," *Animal Behavior,* 3:25–31, 1955.

Regnier, F. E., and Wilson, E. O., "Chemical Communication and 'Propaganda' in Slave-Maker Ants," *Science,* 172:267–69, 1971.

Moulton, D. G., Celebi, G., and Fink, R. P., "Olfaction in Mammals," in Wolstenholme, G. E. W., and Knight, J., eds., *Taste and Smell.* London: J. and A. Churchill, 1970.

Hara, T. J., Ueda, K., and Gorbman, A., "Electroencephalographic Studies of Homing Salmon," *Science,* 149:-884–85, 1966.

Wiener, H., "External Chemical Messengers," I: "Emission and Reception in Man," *New York State Journal of Medicine,* 66:3153–70; II: "Natural History of Schizophrenia," *ibid.,* 67:1144–65.

Smith, K., Thompson, G. F., and Koster, H. D., "Sweat in Schizophrenic Patients: Identification of the Odorous Substance," *Science,* 166:398–99, 1969.

Margolin, A. S., "The Mantle Response of *Diodora aspera,*" *Animal Behavior,* 12:187–94, 1964.

Benacerraf, B., and McDevitt, "The Histocompatibility-Linked Immune Response Genes," *Science,* 175: 273–78, 1972.

Whittaker, R. H., and Feeny, P. P., "Allelochemics: Chemical Interactions Between Species," *Science,* 171:757–70, 1971.

Antaeus in Manhattan

Watson, J. A. L., Nel, J. J. C., and Hewitt, P. H., "Behavioural Changes in Founding Pairs of the Termite *Hodotermes mossambicus,"Journal of Insect Physiology,* 18:373–87, 1972.
Wheeler, W. M., *Essays in Philosophical Biology.* Cambridge, Mass.: Harvard University Press, 1939.
Larousse Encyclopedia of Animal Life. New York: McGraw Hill, 1972.

The Iks

Turnbull, C. M., *The Mountain People.* New York: Simon and Schuster, 1972.

Some Biomythology

Gressitt, J. L., Samuelson, G. A., and Vitt, D. H., "Moss Growing on Living Papuan Moss-Forest Weevils," *Nature,* 217:765, 1968.
Margulis, L., "Symbiosis and Evolution," *Scientific American,* 225(2):48–57, 1971.
Giese, A. C., *Blepharisma: The Biology of a Light-Sensitive Protozoan.* Stanford, Calif.: Stanford University Press, 1973.

On Various Words

Wheeler, W. M. "The Ant-colony as an organism," *Journal of Morphology,* 22:307–25, 1911.

Maeterlinck, M., *The Life of the White Ant.* London: Allen and Unwin, 1930.

Marais, E. N., *Die Siel van die Mier.* Pretoria: J. L. van Schaik, 1933.

Lewis, C. S., *Studies in Words.* Cambridge: Cambridge University Press, 1960.

Morrison, P. "All That Is Made," *Bulletin of the American Academy of Arts and Sciences,* 25(5):7–19, 1972.

Julian of Norwich, *Revelation V,* 1373.

Living Language

Grassé, P. P., "Nouvelles Expériences sur le termite de Muller et considerations sur la théorie de la stigmergie," *Insectes Sociaux,* 14: 73–102, 1967.

Wilson, E. O., *The Insect Societies.* Cambridge, Mass.: Harvard University Press, 1971.